Thermodynamics in Biology

THERMODYNAMICS IN BIOLOGY

Edited by

Enrico Di Cera

UNIVERSITY PRESS
2000

OXFORD
UNIVERSITY PRESS

Oxford   New York

Athens   Auckland   Bangkok   Bogotá   Buenos Aires   Calcutta   Cape Town   Chennai
Dar es Salaam   Delhi   Florence   Hong Kong   Istanbul   Karachi   Kuala Lumpur
Madrid   Melbourne   Mexico City   Mumbai   Nairobi   Paris   São Paulo   Shanghai
Singapore   Taipei   Tokyo   Toronto   Warsaw

and associated companies in
Berlin   Ibadan

Copyright ©2000 by Oxford University Press, Inc.

Published by Oxford University Press, Inc.
198 Madison Avenue, New York, New York 10016

Oxford is a registered trademark of Oxford University Press

All rights reserved. No part of this publication may be reproduced,
stored in a retrieval system, or transmitted, in any form or by any means,
electronic, mechanical, photocopying, recording, or otherwise,
without the prior permission of Oxford University Press.

Library of Congress Cataloging-in-Publication Data
Thermodynamics in Biology / edited by Enrico Di Cera.
    p. cm.
    Includes bibliographical references and index.
    ISBN 0-19-512327-1
    1. Thermodynamics. 2. Biochemistry. I. Di Cera, Enrico.
QP517.T48 T476 2000
572'.01'5367'—dc21   99-086282

9 8 7 6 5 4 3 2 1

Printed in the United States of America
on acid-free paper

Preface

Recent developments in structural biology have advanced our understanding of the molecular basis of biological specificity by revealing the detailed architecture of interfaces for protein–ligand interactions and the precise fold of several proteins and nucleic acids. Recombinant DNA technology has in turn provided the tools to manipulate structures in a site-specific manner, so that precise questions can be asked about the role played by individual amino acids or atoms in stability, folding, or molecular recognition.

At the molecular level, the most prevalent biological functions reduce to a series of alterations in noncovalent and covalent interactions. The current structural emphasis in the study of biological phenomena has reinforced the need for a deeper understanding of the driving forces that determine biological interactions. Because molecular explanations of biological phenomena as inferred from structural information must be informed by and consistent with the laws and principles of thermodynamics, a thorough understanding of biological function requires approaches well balanced between structure and energetics.

This book serves the purpose of defining some state-of-the-art approaches to biological phenomena based on thermodynamics and illustrates the descriptive and predictive power of the laws that govern molecular recognition in biological systems. The book provides key examples of how thermodynamics can assist in the study of biological macromolecules and important phenomena encompassing protein and nucleic acid stability and ligand recognition. Lazaridis and Karplus (chapter 1) offer a microscopic description of the macroscopic thermodynamics of protein folding. Rose and Di Cera (chapter 2) use site-specific thermodynamics to dissect the contribution of individual structural domains to ligand recognition. Marshall, Head, and Ragno (chapter 3) deal with the prediction of the affinity of complexes and thermodynamic functions from structural information. Sharp (chapter 4) illustrates the role of electrostatics in the formulation of thermodynamic properties relevant to molecular recognition. Tinoco and Schmitz (chapter 5) provide a systematic treatment of the thermodynamics of formation of secondary structure in nucleic acids. Mathews, Diamond, and Turner (chapter 6) show how thermodynamics can be applied in the modeling of the secondary structure of RNA. Finally, Bagdassarian and Astumian (chapter 7) cover the topic of the role of conformational fluctuations in protein function and describe the thermodynamics of a motor protein working far from equilibrium.

The breadth of the topics covered in this book illustrates the growing importance of thermodynamic approaches in the study of biological phenomena. As more information continues to emerge from structural studies, and as faster and more accurate computational methods are developed, we will look at biological thermodynamics with renewed interest as a fundamental tool to decipher the rules for specificity and function in proteins and nucleic acids.

Enrico Di Cera

Contents

Contributors   ix

1 Microscopic basis of macromolecular thermodynamics   3
   Themis Lazaridis and Martin Karplus

2 Thermodynamic dissection of cooperativity in ligand recognition   49
   Thierry Rose and Enrico Di Cera

3 Affinity prediction: the sine qua non   87
   Garland R. Marshall, Richard D. Head, and Rino Ragno

4 Electrostatic interactions in proteins and nucleic acids: theory and applications   113
   Kim A. Sharp

5 Thermodynamics of formation of secondary structure in nucleic acids   131
   Ignacio Tinoco, Jr. and Michael Schmitz

6 The application of thermodynamics to the modeling of RNA secondary structure   177
   David H. Mathews, Joshua M. Diamond, and Douglas H. Turner

7 Conformational fluctuations and protein function: the thermodynamics of a Brownian motor   203
   Carey K. Bagdassarian and R. Dean Astumian

Index   227

Contributors

R. Dean Astumian
Departments of Surgery
and Biochemistry and
Molecular Biology
The University of Chicago
Chicago, IL 60637

Carey K. Bagdassarian
The James Frank Institute
The University of Chicago
Chicago, IL 60637

Joshua M. Diamond
Department of Chemistry
University of Rochester
Rochester, NY 14627-0216

Enrico Di Cera
Department of Biochemistry and
Molecular Biophysics
Washington University School
of Medicine
St. Louis, MO 63110

Richard Head
Center for Molecular Design
Washington University School
of Medicine
St. Louis, MO 63110

Martin Karplus
Department of Chemistry and
Chemical Biology
Harvard University
Cambridge,
MA 02138

Themis Lazaridis
Department of Chemistry
City College of New York
New York, NY 10031

Garland R. Marshall
Center for Molecular Design
Washington University School
of Medicine
St. Louis, MO 63110

David H. Mathews
Department of Chemistry
University of Rochester
Rochester, NY 14627-0216

Rino Ragno
Center for Molecular Design
Washington University School
of Medicine
St. Louis, MO 63110

Thierry Rose
Department of Biochemistry
and Molecular Biophysics
Washington University School
of Medicine
St. Louis, MO 63110

Kim. A. Sharp
E. R. Johnson Research Foundation
Department of Biochemistry
and Biophysics
University of Pennsylvania
Philadelphia, PA 19104-6059

Michael Schmitz
Department of Chemistry and
Structural Biology Division
Lawrence Berkeley National
Laboratory
University of California
Berkeley, CA 94720-1460

Ignacio Tinoco, Jr.
Department of Chemistry and
Structural Biology Division
Lawrence Berkeley National
Laboratory
University of California
Berkeley, CA 94720-1460

Douglas H. Turner
Department of Chemistry
University of Rochester
Rochester, NY 14627-0216

Thermodynamics in Biology

1

Microscopic Basis of Macromolecular Thermodynamics

Themis Lazaridis and Martin Karplus

1. The Relevance of Equilibrium Thermodynamics to Biological Phenomena

It has long been assumed that biological systems obey the natural laws of physics and chemistry. This assumption continues to be supported as more details are learned concerning the function of living organisms and the highly complex interactions involved in many essential processes. There is now an enormous amount of information on the events that take place in living systems at the molecular level. However, much of this information is qualitative and descriptive, even when the components involved are known and the structures of many of them (proteins and nucleic acids) have been determined. Many ingenious experiments have been done to establish which phenomena take place, but most of them do not address the question of why things happen the way they do. This is where the physical sciences, including thermodynamics, can make an essential contribution to biology.

Thermodynamics was born in the nineteenth century out of the need to determine the rules governing the efficiency of steam engines, but it rapidly became a general formalism for describing the state and transformations of matter. It is the field of science that provides a direction for the evolution of physical systems. Through the second law, thermodynamics states that a system evolves in the direction that minimizes an appropriate thermodynamic potential, for example the negative of the entropy for isolated systems, or the Gibbs free energy at constant pressure and temperature. It thus generalizes the concept of energy used in mechanics to the concept of free energy by including implicitly the effect of thermal motion and the multiplicity of microscopic states that comprise a given macroscopic state. Application of thermodynamic concepts to the kinetics of chemical reactions (transition-state theory) provides a criteron for selecting the optimal pathway for a transformation, usually the pathway with the transition state of lowest free energy. By providing a direction for systems to evolve and an optimal pathway, thermodynamics offers a formal way for answering *why* things happen the way they do. Thermodynamics is, therefore, an important tool for understanding biological phenomena. However, thermodynamics, which is based on macroscopic measurements, does not provide any direct information concerning the origin of the energy and the entropy changes in chemical or biological processes. This chapter will illuminate the microscopic basis of macromolecular thermodynamics.

Biology has not always had a comfortable relationship with thermodynamics, because of seemingly irreconcilable differences between the simple physical processes traditionally described by thermodynamics and the complex processes that occur in biology. Simple systems rapidly evolve toward equilibrium, which for an isolated system is characterized by maximum entropy (maximum disorder). A simple example is given in fig. 1.1. Upon removal of the barrier, the fluids rapidly reach a new equilibrium state with uniform distribution of the two components in the whole system. In contrast, living systems never reach equilibrium and in many cases evolve toward states of increasing order. For example, biological development involves the growth of a complex multicellular organism out of a single cell. A complete thermodynamic analysis of such a complex order-generating process is a challenge to thermodynamicists (Zotin, 1972). It is easy to argue that such phenomena do not contradict the laws of thermodynamics, simply because biological systems are open systems far from equilibrium and, although order is generated in the biological system, the entropy of the universe as a whole may still increase (Zotin, 1972). However, this argument shows only that biological ordering can arise (which we already know by experience); it does not explain why it occurs. Why out of all possible ways of increasing the entropy of the universe does nature choose one that involves creation of order in a few of its subsystems? A complete analysis of biological phenomena in terms of thermodynamics does not yet exist. Some interesting studies of evolutionary mechanisms and the appearance of order in biological systems have been made both experimentally and theoretically (Eigen, 1986; Kauffman, 1983; Arnold, 1998).

The question of how biological systems evade equilibrium was taken up by Schrödinger (1944) in his influential treatise "What is life?". He proposed that living organisms feed on negative entropy which they import from their surroundings. Actually, import of energy in an appropriate form (other than thermal) can also be used to maintain a system far from equilibrium. The intake of solar energy by the photosynthetic apparatus of plants provides the motive power for life on earth. This energy gradually "trickles" down to other parts of the plant, and from plants to animals through the food cycle. The efficiency of the whole process is high because the energy is maintained in the form of chemical energy;

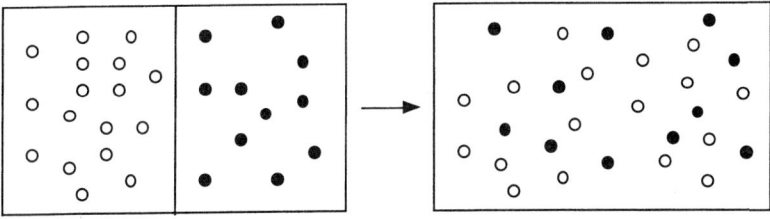

Figure 1.1. A simple process evolving toward equilibrium

that is, it does not generate much unrecoverable thermal energy. At a more microscopic level, it is likely that all individual processes in the cell evolve toward equilibrium; but equilibrium is not reached, because the external conditions change and shift the equilibrium point to some other state. For example, the folded conformation of a protein may be the lowest-free-energy state at neutral pH, and indeed the protein evolves toward the folded state at neutral pH. If at some future point the pH is changed, the protein will evolve toward its new equilibrium state, which may be the unfolded state. In fact, differences in pH in different parts of an organism play an important role in biological processes, including the entrance of viruses into cells (Wiley and Skehel, 1987). That biological macromolecules have marginal stability is in part due to their need to respond readily to changes in the environmental conditions. It should be noted that the lifetime of individual protein molecules is generally much shorter than that of the organism. Apparently, processes in the cell are dynamically coupled in such a way that overall equilibrium is never reached in a living system. An analogous situation for the system in fig 1.1 would arise if one of the two particle types were magnetic, and if a magnet would move close to and far from the box, causing phase separation when it is close and allowing mixing when distant. The goal of biological thermodynamics is to "explain" biological processes with the same rigor and clarity as the simple mixing–unmixing paradigm of fig. 1.1.

Thermodynamics is most fully developed for equilibrium states. In fact, the word *thermodynamics* is most often used to mean *equilibrium thermodynamics*. The description of nonequilibrium states involves a larger number of independent variables, some of which are not simple scalars like temperature or pressure, but are vectors or tensors. This increases the level of mathematical complexity. The so called "linear thermodynamics of irreversible processes" (Prigogine, 1961; de Groot and Mazur, 1984) has extended the scope of thermodynamics to systems near equilibrium. This field has produced some useful results, and some important applications to biological systems have been made (Katchalsky and Curran, 1965; Jou and Llebot, 1990). However, from a practical viewpoint, it does not seem to offer dramatic advantages over the traditional continuum-mechanical treatment of transport phenomena. The theory of *dissipative structures* (Glandsorff and Prigogine, 1971) addresses the problem of how order can arise spontaneously in systems far from equilibrium. A detailed analysis of biological structure formation at the molecular level in the light of this theory still remains to be achieved. Nonequilibrium thermodynamics is still under development, and its full potential for applications to biology has not yet been realized. In what follows in this chapter, we focus on equilibrium thermodynamics and its microscopic correlates.

Despite the fact that biological processes do not occur under equilibrium conditions, equilibrium thermodynamics has been extensively used in biochemistry. In the usual reductionistic fashion, biochemists have isolated specific processes and studied them in vitro under equilibrium conditions. One reason for this is that the measurements are much easier under equilibrium conditions. One example of a process that has been subjected to detailed biophysical characterization is protein folding. It was shown by Anfinsen (Anfinsen, 1973) that ribonuclease A folds spontaneously to its native state in vitro. The experiments suggested that such processes (or at least this process, protein folding) do evolve towards equilibrium in the cell and are similar, therefore, to the simple physical processes studied by traditional thermodynamics. Changes inside the cell which are uphill in free energy are coupled to others that are downhill in free energy (usually hydrolysis of ATP), so that the overall process is spontaneous (Eisenberg and Crothers, 1979).

Because of its importance, the equilibrium-thermodynamic properties of protein folding have been subjected to detailed measurements; that is, the free energy, enthalpy, entropy, and heat capacity have been measured (Brandts, 1964; Privalov, 1979; Velicelebi and Sturtevant, 1979; Makhatadze and Privalov, 1995). Corresponding detailed studies in cells are much more difficult, and it is assumed that the thermodynamics of protein folding in the cell is essentially the same as it is in solution. However, it should be noted that the cellular medium is very different from the solution conditions used in most measurements, so that non-negligible differences in behavior can occur (Ellis and Hartl, 1999). Another common application of thermodynamics is to binding of ligands to biomolecules. Spectroscopic methods provide the equilibrium constant (free energy of binding), and isothermal titration calorimetry provides the enthalpy and heat capacity of binding (Fisher and Singh, 1995). These thermodynamic quantities are useful for interpreting the mechanism at the molecular level. For example, very negative entropies and heat capacities of binding have been used to suggest that the binding partners are flexible in isolation and become ordered upon binding (Spolar and Record, 1994). However, interpretation of thermodynamic quantities in terms of a detailed molecular mechanism is not straightforward, because

MACROMOLECULAR THERMODYNAMICS 7

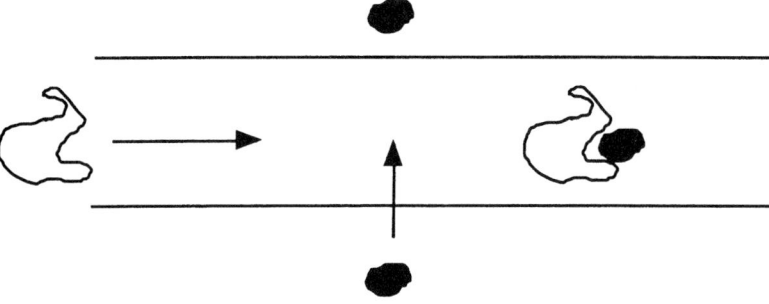

Figure 1.2. Nonequilibrium ligand binding

usually more than one molecular mechanism is consistent with the macroscopic data. This is an area where molecular modeling and statistical-mechanical theories can play an important role.

Equilibrium-thermodynamic measurements can be very useful and relevant to the situation in vivo, because—despite the lack of overall equilibrium in the cell—there can exist partial equilibrium, either temporal (with a certain time scale) or spatial (within a certain region). Nevertheless, one has to keep in mind the limitations of such measurements. One hypothetical example is depicted in fig. 1.2. A protein flows in a channel, and a ligand crosses the channel and has the opportunity to bind to the protein. It is of interest to know what percentage of ligand is taken up by the protein, and what percentage of the protein binds ligand. Measurement of the equilibrium binding constant will be useful only if the kinetics of binding is fast relative to the rate of transport of the protein and the ligand. Otherwise, it will be completely irrelevant, and a kinetic analysis would be required.

Classical thermodynamics is a macroscopic science which is not concerned with the microscopic nature of matter. Since biological phenomena depend critically on the nature of the molecules involved, thermodynamics is most useful in biology when combined with a microscopic description. Molecular models that describe the interactions within and between molecules are critically important. The link between microscopic interactions and macroscopic properties is provided by statistical mechanics. Statistical mechanics considers an ensemble of microscopic systems consistent with a given set of macroscopic conditions. The thermodynamic internal energy is obtained as a Boltzmann-weighted average of the energy of all microscopic states. The entropy and free energy are obtained from the partition function (the sum of the Boltzmann factors of the microscopic states). Direct estimation of the partition function is possible only for very simple systems. For realistic systems, approximations need to be made. In this chapter, we illustrate the microscopic approaches by focusing on the thermodynamics of protein folding and its interpretation in terms of the interactions involved. We start with a theoretical framework for the stability of macromolecules that provides information concerning the interactions that contribute to the free energy of folding. These are analyzed in subsequent sections.

2. Stability of Macromolecular Conformations

Biopolymers, such as proteins and nucleic acids, adopt essentially unique conformations under physiological conditions. One central question is whether this conformation is under thermodynamic or kinetic control— that is, whether the native protein conformation corresponds to the most stable (thermodynamic control) or to the kinetically most accessible (kinetic control) conformation. The first situation would arise if the barriers for interconversion of conformations were small enough to be traversed within experimental time scales. Thus, equilibrium is established and the conformations are populated according to the Boltzmann distribution. On the other hand, if the barriers between conformations are too high, then the system is no longer ergodic, and the macromolecule settles in the lowest local minimum it can find within the available time. There has been no proof of either of the two situations. The original experiments of Anfinsen seem to support the thermodynamic hypothesis. Recent experiments, however, have suggested that there may be exceptions, especially for larger and more complex proteins (Goldberg, 1985; Baker and Agard, 1994; Baker, 1998). It is also of interest to mention that such metastability has been found in a highly simplified, but detailed, simulation model of protein folding (Dinner and Karplus, 1998). It is very likely that kinetic control is even more prevalent in complex cellular processes. For example, the migration of lipids from one leaflet of a lipid bilayer to the other can take days, so that a nonequilibrium composition in biological membranes can be maintained throughout the lifetime of a cell (Jain, 1988).

Equilibrium-thermodynamic concepts have an important role even if the thermodynamic hypothesis is not valid in all cases, because partial equilibrium can exist with respect to certain degrees of freedom even when overall equilibrium is not established. One example concerns the degrees of freedom for the solvent. For most changes of macromolecular conformations, the solvent equilibrates on a picosecond timescale (Halle et al., 1981). For configurations where water needs to equilibrate between the bulk and an internal cavity, equilibration can take much longer (Otting et al., 1991; Ernst et al., 1995), but rarely longer than the experimental time scales of seconds or minutes. Because equilibration of the solvent is rapid in comparison with movements of the macromolecule, the solvent degrees of freedom can be integrated out to give the equilibrium solvation free energy. This quantity is then added to the internal macromolecular energy to give the "effective energy" or "potential of mean force" for each macromolecular conformation. The effective energy defines a hypersurface in the conformational space of the molecule (the "energy landscape" (Frauenfelder et al., 1991)) whose shape determines the conformational properties of the macromolecule, independently of whether it is under thermodynamic control with respect to the macromolecular degrees of freedom.

Formal integration of the solvent's degrees of freedom can be accomplished using standard statistical-mechanical methods (McQuarrie, 1976). We briefly review what is involved, using the notation of Lazaridis and Karplus (1999b). Consider a macromolecule consisting of M atoms with Cartesian coordinates $\mathbf{R}_i = (X_i, Y_i, Z_i)$ ($i = 1, \ldots, M$) and internal coordinates q_i ($i = 1, \ldots, 3M - 6$). The macromolecule is immersed in a bath of N rigid solvent molecules with coordinates $r_i = (x_i, y_i, z_i, \omega_i, \varphi_i, \chi_i)$ ($i = 1, \ldots, N$), where x, y, and z are Cartesian coordinates of the center of mass and ω, φ, and χ are the Euler angles specifying the orientation. For simplicity we assume constant temperature and volume conditions corresponding to the canonical ensemble. The canonical partition function is

$$Q = \frac{Z}{N! \Lambda^{3M} \Lambda^{3N}}, \tag{1}$$

where Λ is the de Broglie wavelength and Z is the classical configurational integral

$$Z = \int \exp(-\beta H) d\mathbf{r}^N d\mathbf{R}^M, \tag{2}$$

with H the Hamiltonian and $\beta = 1/kT$. The Helmholtz free energy is given by

$$A = -kT \ln Q = -kT \ln Z + kT \ln(N! \Lambda^{3M} \Lambda^{3N}). \tag{3}$$

One can formally perform the integration over the solvent coordinates in eq. 1 through defining the potential of mean force, W, by

$$\exp(-\beta W) = Z_{ww}^{-1} \int \exp(-\beta H) \, d\mathbf{r}, \tag{4}$$

where

$$Z_{ww} = \int \exp(-\beta H_{ww}) \, d\mathbf{r} \tag{5}$$

is the pure-solvent configurational integral. By introducing W, the configurational integral can be written

$$Z = Z_{ww} \int \exp(-\beta W) \, d\mathbf{R}. \tag{6}$$

Thus, the integral in eq. (6) depends explicitly only on the macromolecular degrees of freedom $\mathbf{R}$. If the Hamiltonian is additive, as it is in most molecular-mechanical force fields (e.g., MacKerell et al., 1998), further explicit simplifications are possible. We can write H as $H = H_{mm} + H_{mw} + H_{ww}$, where the three components are the intramacromolecular, macromolecule–solvent, and solvent–solvent interactions, respectively. We then obtain

$$\exp(-\beta W) = \exp(-\beta H_{mm}) Z_{ww}^{-1} \int \exp(-\beta H_{mw} - \beta H_{ww}) \, d\mathbf{r}^N, \qquad (7)$$

so that

$$W = H_{mm} + X, \qquad (8)$$

where

$$\exp(-\beta X) = Z_{ww}^{-1} \int \exp(-\beta H_{mw} - \beta H_{ww}) \, d\mathbf{r}^N. \qquad (9)$$

Equation (9) for X can be written as

$$X = -kT \ln \langle \exp(-\beta H_{mw}) \rangle_o \equiv \Delta G^{slv}, \qquad (10)$$

where the ensemble average $\langle \cdot \rangle_o$ is taken over the pure solvent. Equation (10) is the familiar expression for the excess chemical potential, or standard solvation free energy (Ben-Naim, 1978). Thus, the potential of mean force consists of two terms: the intramacromolecular energy and the solvation free energy. Instead of the term *potential of mean force*, we can use the more intuitive term *effective energy*. The function W defines a hypersurface in the conformation space of the macromolecule in the presence of equilibrated solvent (the so-called energy landscape). It determines the thermodynamics and kinetics of macromolecular conformational transitions.

This separation of the effective energy can be accomplished formally in the case of nonpairwise additive potentials. For example, in the presence of three-body forces, the Hamiltonian can be written.

$$H = (H_{mm} + H_{mmm}) + (H_{mw} + H_{ww} + H_{www} + H_{wwm} + H_{mmw}), \qquad (11)$$

and the effective energy would be

$$W = (H_{mm} + H_{mmm}) + -kT \ln \langle \exp(-\beta H_{mw} - \beta H_{mmw} - \beta H_{mww}) \rangle_o. \qquad (12)$$

However, for this case, the actual evaluation of W is much more difficult.

For a formal description of the macromolecule, it is often more convenient to use the internal coordinates $\mathbf{q} = (q_i : i = 1, \ldots, 3M - 6)$. The Jacobian for the transformation depends only on bond lengths and bond angles (Herschbach et al., 1959; DeVoe, 1969) and is, therefore, approximately constant for all conformations and can be taken out of the integral in eq. 6. Here we include it in the notation $d\mathbf{q}$. The integration over the six external coordinates can be performed since the system is homogeneous, to give $8\pi^2 V$ so that

$$Z = 8\pi^2 Z_{ww} V \int \exp(-\beta W)\, d\mathbf{q}. \tag{13}$$

It can be shown that the probability of finding the system at the configuration $\mathbf{q}$ is (Lazaridis and Karplus, 1999b)

$$p(\mathbf{q}) = \frac{\exp[-\beta W(\mathbf{q})]}{\int \exp[-\beta W(\mathbf{q})]\, d\mathbf{q}}. \tag{14}$$

Consequently,

$$\int p(\mathbf{q}) \ln p(\mathbf{q})\, d\mathbf{q} = \int p(\mathbf{q})[-\beta W(\mathbf{q}) - \ln Z + \ln Z_{ww} + \ln 8\pi^2 V]\, d\mathbf{q}$$
$$= -\ln Z + \ln Z_{ww} + \ln 8\pi^2 V - \beta \int p(\mathbf{q}) W(\mathbf{q})\, d\mathbf{q}, \tag{15}$$

and, from eq. (3),

$$A = A^\circ + kT \ln \frac{\Lambda^{3M}}{8\pi^2 V} + \int p(\mathbf{q})[H_{mm}(\mathbf{q}) + \Delta G^{slv}(\mathbf{q})]\, d\mathbf{q} + kT \int p(\mathbf{q}) \ln p(\mathbf{q})\, d\mathbf{q}$$
$$= A^\circ + kT \ln \frac{\Lambda^{3M}}{8\pi^2 V} + \langle W \rangle - TS^{conf}, \tag{16}$$

where A° is the free energy of the pure solvent, and where the second term is the ideal contribution from macromolecular translation and rotation. The third term in eq. (16) is the average effective energy, which is equal to the average intramolecular energy plus the average solvation free energy. The last term is the contribution of the configurational entropy of the macromolecule to the free energy. The solvent entropy is contained in $\Delta G^{slv}(\mathbf{q})$ and in $p(\mathbf{q})$. The Gibbs free energy is equal to the Helmholtz free energy plus the PV term. Under ambient conditions, the PV term is negligible, and Gibbs and Helmholtz free energies can be used interchangably; we do so in this chapter.

Expressions can also be obtained for the energy and entropy of the system. The energy has the form

$$E = kT^2 \left(\frac{\partial \ln Q}{\partial T}\right)_{N,V}, \tag{17}$$

which gives

$$E = \tfrac{3}{2} kT(M + N) + \int p(\mathbf{q}, \mathbf{r}^N) H(\mathbf{q}, \mathbf{r}^N)\, d\mathbf{q}\, d\mathbf{r}^N, \tag{18}$$

where the first term is the kinetic energy and the second the potential energy. The conditional probability distribution $p(\mathbf{r}|\mathbf{q})$ of finding the solvent configuration $\mathbf{r}$, given that the macromolecule is in conformation $\mathbf{q}$, is defined by

$$p(\mathbf{q}, \mathbf{r}^N) = p(\mathbf{q}) p(\mathbf{r}^N | \mathbf{q}). \tag{19}$$

With eqs. (19) and (6) we obtain for the potential energy

$$\int p(\boldsymbol{q})H_{\text{mm}}(\boldsymbol{q})\,d\boldsymbol{q} + \int p(\boldsymbol{q})\,d\boldsymbol{q} \int p(\boldsymbol{r}^N|\boldsymbol{q})[H_{\text{mw}}(\boldsymbol{q},\boldsymbol{r}^N) + H_{\text{ww}}(\boldsymbol{r}^N)]\,d\boldsymbol{r}^N, \quad (20)$$

The first term in eq. (20) is the average intramolecular energy and the second the average solute-solvent and solvent-solvent energy.

The entropy is given by

$$S = k\ln Q + kT\left(\frac{\partial \ln Q}{\partial T}\right)_{N,V}, \quad (21)$$

which yields

$$\begin{aligned}
S &= -k\int p(\boldsymbol{q},\boldsymbol{r}^N)\ln p(\boldsymbol{q},\boldsymbol{r}^N)\,d\boldsymbol{q}\,d\boldsymbol{r}^N - k\ln(N!\Lambda^{3M}\Lambda^{3N}) + \left(\frac{3}{2}\right)k(M+N) \\
&= -k\int p(\boldsymbol{q})\ln p(\boldsymbol{q})\,d\boldsymbol{q} - k\int p(\boldsymbol{q})\,d\boldsymbol{q}\int p(\boldsymbol{r}|\boldsymbol{q})\ln p(\boldsymbol{r}|\boldsymbol{q})\,d\boldsymbol{r} \\
&\quad - k\ln(N!\Lambda^{3M}\Lambda^{3N}) + \frac{3}{2}k(M+N),
\end{aligned} \quad (22)$$

where the first term is the configurational entropy of the macromolecule and the second term is the average solvent entropy—i.e., the entropy that arises from solute–solvent and solvent–solvent correlations.

By use of the above analysis, it is possible to provide a more detailed microscopic description of the native state of a protein, given that it is under thermodynamic control. The probability distribution $p(\boldsymbol{q})$ completely specifies the conformational *state* of a protein. The native state could be defined as the distribution of configurations of the macromolecule, $p(\boldsymbol{q})$, that minimizes the free-energy functional of eq. (16) under physiological conditions; a variational minimizatioan of eq. (16) gives eq. (14). This definition has the advantage of including protein flexibility and accounting for possible disorder in the native state. For convenience, the native state is usually defined as including only certain values of $\boldsymbol{q}$ (see below). The average-effective-energy term tends to localize the macromolecule in the deepest wells of the multidimensional effective-energy surface (Elber and Karplus, 1987), but the configurational entropy term tends to make $p(\boldsymbol{q})$ as uniform as possible. As a result, the native state, which consists of the conformations of lowest free energy, need not be the ones of lowest effective energy, since some deep wells may be so narrow that the vibrational entropy of a protein in those wells would be very small.

Integration of the solvent's degrees of freedom greatly simplifies the treatment of conformational equilibria. However, this remains only a formal exercise unless a model for the solvation free energy as a function of macromolecular conformation is available. The derivation of such models is an active area of research. It can, in principle, be done by using statistical thermodynamics, but it has so far been based more on intuitive arguments because of theoretical difficulties. The statistical thermodynamics of solvation is considered in section 4, and empirical solvation models are discussed in section 5. In section 6, we present a simplified theoretical model that is parameterized according to experimental data.

The thermodynamic stability of a macromolecular native state can be expressed in terms of the standard free energy of folding, ΔG. Given the equilibrium constant K for the folding reaction, we have

$$\Delta G = -RT \ln K, \qquad K = \frac{[\text{native}]}{[\text{denatured}]}. \tag{23}$$

This equilibrium constant cannot be measured under physiological conditions, because the concentration of the denatured state is vanishingly small. Consequently, K and ΔG are usually determined either at high temperature or at high denaturant concentration, and the results are extrapolated to physiological conditions (room temperature, or zero denaturant concentration). Under usual physiological conditions, the Gibbs free energy is essentially equal to the Helmholtz free energy; the $P\Delta V$ term is negligible. We use ΔG because most measurements are made under constant pressure conditions.

With statistical mechanics, we can derive an expression for the free energy of folding in terms of the interactions and the distributions of microscopic states. We divide the configurational space into subsets $\mathcal{A}$ consisting of different configurations for the macromolecule in analogy to the treatment of the equilibrium between different isomers. The free energy of conformational set $\mathcal{A}$ is

$$A_{\mathcal{A}} = -kT \ln Z_{\mathcal{A}} + kT \ln(N! \Lambda^{3M} \Lambda^{3N}), \tag{24}$$

with

$$Z_{\mathcal{A}} = 8V\pi^2 Z_{\text{ww}} \int_{\mathcal{A}} \exp(-\beta W)\, d\mathbf{q}, \tag{25}$$

where the integration is carried out over the configurations in set $\mathcal{A}$. By definition,

$$\sum_{\mathcal{A}} Z_{\mathcal{A}} = Z. \tag{26}$$

The free energy of the conformational set $\mathcal{A}$ can then be written:

$$\begin{aligned}A_{\mathcal{A}} &= A^{\circ} + kT \ln \frac{\Lambda^{3M}}{8\pi^2 V} + \int_{\mathcal{A}} p_{\mathcal{A}}(\mathbf{q}) W(\mathbf{q})\, d\mathbf{q} + kT \int_{\mathcal{A}} p_{\mathcal{A}}(\mathbf{q}) \ln p_{\mathcal{A}}(\mathbf{q})\, d\mathbf{q} \\ &= A^{\circ} + A^{\text{id}} + \langle W \rangle_{\mathcal{A}} - T S_{\mathcal{A}}^{\text{conf}},\end{aligned} \tag{27}$$

where $p_{\mathcal{A}}$ is a probability distribution normalized within the set A; that is,

$$p_A(q) = \frac{\exp[-\beta W(q)]}{\int_A \exp[-\beta W(q)]\,dq}. \tag{28}$$

The first two terms in eq. (27) are, as in eq. (16), the free energy of pure solvent and the ideal translational and rotational free energy of the macromolecule. These are the same for all conformational states. The third term in eq. (27) is the average effective energy of state $\mathcal{A}$ and the last term is the configurational entropy of state $\mathcal{A}$. The free-energy difference between the two sets $\mathcal{A}$ and $\mathcal{B}$ is then

$$\Delta A = A_B - A_A = A[p_B(q)] - A[p_A(q)] = \langle W \rangle_B - \langle W \rangle_A - T(S_B^{\text{conf}} - S_A^{\text{conf}}) = \Delta\langle W \rangle - T\Delta S^{\text{conf}} = \Delta\langle H_{\text{mm}} \rangle + \Delta\langle \Delta G^{\text{slv}} \rangle - T\Delta S^{\text{conf}}, \tag{29}$$

where the notation $A[p(q)]$ denotes that the free energy is a functional of the distribution function. One can also use $A[p_B(q)] - A[p_A(q)]$ with p_A and p_B defined as the conformational distributions under different external conditions—for example, one under physiological conditions and another for high-denaturant concentrations. This is more consistent with the definition of the native state given above, and does not require an arbitrary separation of the conformational space into "native" and "denatured" regions. The difference in Gibbs free energy between $\mathcal{A}$ and $\mathcal{B}$ is obtained by adding the $P\Delta V$ term: that is,

$$\Delta G = \Delta\langle H_{\text{mm}} \rangle + \Delta\langle \Delta G^{\text{slv}} \rangle - T\Delta S^{\text{conf}} + P\Delta V \tag{30}$$

If $\mathcal{A}$ is the denatured state and $\mathcal{B}$ the native state, both of which have to be defined in some way and both of which include many configurations, then eq. 29 gives the free energy of folding. This equation expresses the intuitive idea that protein stability is a result of a balance between the effective energy, which favors the native state, and the configurational entropy, which favors the denatured state. The change in average effective energy is related to the depth of the native-state well on the effective-energy hypersurface, though there may be a barrier between the two states. The entropic cost of localizing the protein in this well was estimated to be of the order of a few hundred kcal/mol (Kauzmann, 1959). Since the overall free energy change upon folding is usually between 5 and 15 kcal/mol (Makhatadze and Privalov, 1995), the depth of the native-state well on the effective energy surface is also of the order of a few hundred kcal/mol (Lazaridis et al., 1995).

3. Energy Functions

An essential element in developing a microscopic description for the interpretation of protein thermodynamics is the potential-energy function, which makes it possible to calculate the potential energy of the system as a function of the atomic coordinates. The potential energy can be used directly to determine the relative stabilities of the different possible structures of the system. To obtain the forces acting on the atoms of the system, the first derivatives of the potential with respect to the atoms' positions are calculated. These forces can be used to determine dynamic and thermodynamic properties of the system—that is, by solving Newton's equations of motion to describe how the atomic positions change with respect to time and by calculating average properties, such as the enthalpy, from these positions (McCammon and Harvey, 1987; Brooks, et al., 1988). From the second derivatives of the potential surface, the force constants for small displacements can be evaluated and used to find the normal modes. The normal modes provide an alternative approach to the dynamics in the harmonic limit. They are very useful also for introducing quantum corrections to the vibrational contributions to thermodynamic quantities.

To obtain potential energy surfaces for proteins with the required accuracy and speed, it is necessary to introduce a simple model which is calibrated by fitting it to experimental or quantum-mechanical information. When working with macromolecules, there is a need to have available a reliable method for calculating interaction energies many times (10^4 to 10^6 energy calculations) for systems of hundreds to thousands of atoms. Such a method is supplied by empirical energy functions. However, there is a price to pay for introducing this type of model for the calculation. Empirical energy functions do not have the generality of quantum-mechanical calculations. They are at best limited to the systems for which they were designed.

The potential energy, $U(\mathbf{R})$ of the macromolecule, as a function of the atomic coordinates $\mathbf{R}$, has the form

$$U(\mathbf{R}) = \sum_{\text{bonds}} K_b(b-b_0)^2 + \sum_{\text{UB}} K_{\text{UB}}(S-S_0)^2 + \sum_{\text{angle}} K_\theta(\theta-\theta_0)^2 +$$
$$\sum_{\text{dihedrals}} K_\chi[1+\cos(n\chi-\delta)] + \sum_{\text{impropers}} K_{\text{imp}}(\phi-\phi_0)^2 + \qquad (31)$$
$$\sum_{\text{nonbond}} \varepsilon_{ij}\left[\left(\frac{R_{\text{min}ij}}{r_{ij}}\right)^{12} - 2\left(\frac{R_{\text{min}ij}}{r_{ij}}\right)^6\right] + \frac{q_i q_j}{\varepsilon r_{ij}},$$

where K_b, K_{UB}, K_θ, K_χ, and K_{imp} are the bond, Urey–Bradley, angle, dihedral-angle, and improper-dihedral-angle force constants, respectively; b, S, θ, χ, and ϕ are the bond length, Urey–Bradley (1,3) distance, bond angle, dihedral angle, and improper torsion angle, respectively (all the internal coordinates are expressed as functions of $\mathbf{R}$); the subscript zero represents the values for which the individual terms have their minima. The dihedral term depends on the parameters n and δ, which define the multiplicity (or periodicity) and phase, respectively. Coulomb and Lennard–Jones 6–12 terms make up the external or nonbonded interactions; ε_{ij} is the Lennard–Jones well-depth and $R_{\text{min}\,ij}$ is the distance at the Lennard–Jones minimum; q_i is the partial atomic charge, ε is the effective dielectric constant, and r_{ij} is the distance between atoms i and j, respectively. The Lennard–Jones parameters between pairs of different atoms are obtained from the Lorentz–Berthelot combination rules, in which ε_{ij} and $R_{\min ij}$ are derived as the geometric mean and the arithmetic mean, respectively, of the parameters for atoms i and j.

As is evident from eq. (31) the potential energy function is simple in form. This simplicity is achieved, with little sacrifice in accuracy, for the properties of primary interest for the macromolecules for which the potential function is designed. The bond and angle energies are treated with harmonic terms, as is the Urey–Bradley term. Thus, the making and breaking of bonds cannot be treated directly using the standard potential-energy function. However, it is possible to study such processes (e.g. chemical reactions) by use of quantum-mechanical/molecular-mechanical (QM/MM) potential-energy functions (Field et al., 1990). Use of the harmonic terms is satisfactory for the majority of condensed-phase simulations, which are performed close to or below room temperature, such that configurations with large deviations from the minimum-energy bond lengths and angles are usually unimportant. Even high-temperature simulations (e.g. protein unfolding at 600 K) have been shown to yield meaningful results without modification of the potential function (Caflisch and Karplus, 1995). Rotations about bonds are treated with a sinusoidal function, which may involve a Fourier series, of the length needed for the accurate treatment of torsional surfaces. In most cases only one or two terms in the series are required to obtain results of sufficient accuracy. For each pair of bonded atoms, terms for all possible definitions of the dihedral angle are used. Improper dihedral terms, traditionally used to maintain chirality in extended atomic potential functions, are included to allow for fine tuning of specific properties, such as ring-torsional deformations and out-of-plane bending of aromatic hydrogen atoms. Simple Coulombic and Lennard–Jones terms are used for the nonbonded interactions. The latter are important because of the central role of nonbonded interactions in macromolecular structure and dynamics and the computational costs associated with the calculation of these terms.

Based on ab initio calculations (Reiher, 1985), no explicit hydrogen-bonding terms are used. Instead, it was found that the Coulomb and Lennard–Jones terms in the potential function provided an accurate representation of hydrogen-bonding interactions; see also Hagler et al. (1974). We believe that the quality of this potential function is such that it will continue to be used for many simulations of macromolecular systems. Recent examples are the determination of the free-energy surface for the folding of a protein as a function of the radius of gyration in the presence of explicit solvent (Boczko and Brooks, 1995) and its implementation with implicit solvent corrections and use in high-temperature unfolding studies (EEF1, Lazaridis and Karplus, 1997) or peptide-folding equilibria at room temperature (ACE, Schaefer and Karplus, 1996) (see section S5,6).

The parameters were determined by optimizing the reproduction of experimental pure-liquid, solution, and crystal data, as well as the appropriate ab initio results. A particular effort was made in the optimization of the partial atomic charges and Lennard–Jones parameters. The charges were chosen to be consistent with the CHARMM version (Reiher, 1985; Neria et al., 1996) of the TIP3P water model (Jorgensen et al., 1983), and the Lennard–Jones parameters were based primarily on pure-solvent properties. Selection of the TIP3P water model dictated the solvent–solvent interactions in the force field. Optimization of the solute (e.g. protein, nucleic acid, or lipid) interaction parameters focused on balancing the solvent–solute and solute–solute interactions with respect to the solvent–solvent interactions. This was achieved by comparison with ab initio calculations for the corresponding interactions in various dimer systems (Reiher, 1985). More recently, all-atom potential functions have been developed that also take account of fitting solid state and liquid properties. Because of the importance of empirical potential functions as the basis of statistical-mechanical treatment of macromolecules, continued effort to refine these functions continue in many laboratories. However, the presently available energy functions or force fields are clearly accurate enough for many semi-quantitative applications, including those described in this chapter (MacKerell, et al., 1998).

4. Statistical Thermodynamics of Solvation

In molecular dynamics simulations, solvation is usually treated by surrounding the macromolecule with a large number of explicit water molecules. This approach has two main limitations. First, the computational expense is exceedingly high: most CPU time in the simulation is expended calculating the motions of the solvent molecules, which often are of no direct interest. The second limitation is that, in explicit solvent simulations, the effective energy of a macromolecular conformation is not known; only the intramolecular energy is known. The solute–solvent and solvent–solvent energies can be calculated as well, but they are not directly related to the solvation free energy. The alternative to explicit solvation is to include in the energy function a model for the solvation free energy—that is, to perform simulations with an *effective* energy function. This approach is referred to as implicit solvation, and is about two orders of magnitude faster than corresponding simulations with explicit solvent.

The solvation free energy in eq. 10 is more traditionally known as the excess chemical potential. It is the part of the chemical potential that depends on the interactions between the solute and the solvent (it is zero for ideal gas particles). It can be thought of as the free energy of transferring the solute from a fixed point in the gas phase to a fixed point in the solution (Ben-Naim, 1978). Progress in the theoretical study of the conformational properties of macromolecules depends critically on development of quantitative models for the solvation free energy for these systems.

Statistical thermodynamics has been used to develop several different approaches for calculating the solvation free energy. Analytical theories are applicable only to simple fluids, such as hard spheres and Lennard–Jones particles. For example, the thermodynamic properties of hard-sphere fluids are quite well predicted by scaled-particle theory (Reiss, 1965) and by integral-equation theories (particularly, the Percus–Yevick approximation) (Hansen and McDonald, 1986). For aqueous solutions, the XRISM integral equation (Hirata and Rossky, 1981; Pettitt and Rossky, 1986; Yu et al., 1991) has been used to obtain the thermodynamic solvation properties of small solutes. The predictions of XRISM are qualitatively correct but suffer from quantitative deficiencies (Ichiye and Chandler, 1988; Yu et al., 1991). Solvation free energies for complex systems, or more specifically small changes in complex systems (e.g. the effect of an amino-acid mutation of a protein), can be calculated from molecular dynamics and Monte Carlo simulations, combined together with techniques, such as free-energy perturbation theory and thermodynamic integration methods (Beveridge and DiCapua, 1989; Kollman, 1993). Such methods are exact, in principle, for given intermolecular potentials, but in practice they usually require very long configurational sampling times to converge.

Other than quantitative deficiencies, the major limitation of these methods is that they do not provide a strategy for going from small molecules to macromolecules. A possible exception is the theory for the hydrophobic solvation recently developed by Lum et al. (1999); its range of practical applicability has still to be determined. Performing solvation free-energy calculations for every single conformation of a macromolecule that we want to consider is out of the question. Even integral-equation theories, which are computationally much more efficient than free-energy simulations, require the numerical solution of an integrodifferential equation. This is too expensive to do at every step of a molecular dynamics run. What is needed is simple semi-analytical models that describe the solvation free energy of macromolecules as a function of conformation. Ideally, the results of rigorous calculations of solvation free energy should be used as the basis for building such models for macromolecules. Lacking such results, experimental data can serve as a basis for constructing models for the solvation free energy.

One approach that holds promise in this regard is the *inhomogeneous-fluid* solvation theory, so named because it views the solutes as inhomogeneities in the solvent and treats the solution as an inhomogeneous system (Matubayasi et al., 1994; Lazaridis, 1998a). This approach considers the energetic and entropic contributions separately, namely

$$\Delta G^{\text{slv}} = \Delta E^{\text{slv}} - T\Delta S^{\text{slv}} + P\Delta V^{\text{slv}}, \tag{32}$$

and gives the solvation energy and entropy as a sum of a solute–solvent and a solvent–solvent term; that is,

$$\Delta E^{\text{slv}} = E_{\text{sw}} + \Delta E_{\text{ww}}, \tag{33}$$

$$\Delta S^{\text{slv}} = S_{\text{sw}} + \Delta S_{\text{ww}}. \tag{34}$$

For solute insertion at a fixed point in the solvent (Ben-Naim, 1978), we have (Lazaridis, 1998a)

$$E_{\text{sw}} = \rho \int g^{(1)}(\mathbf{r}) u_{\text{sw}}(\mathbf{r})\, d\mathbf{r}, \tag{35}$$

$$\Delta E_{\text{ww}} = \tfrac{1}{2}\rho^2 \int g^{(1)}(\mathbf{r})[g^{(1)}(\mathbf{r}') - 1]g^{(2)}(\mathbf{r},\mathbf{r}') u_{\text{ww}}(\mathbf{r},\mathbf{r}')\, d\mathbf{r}\, d\mathbf{r}', \tag{36}$$

$$S_{\text{sw}} = -k\rho \int g^{(1)}(\mathbf{r}) \ln g^{(1)}(\mathbf{r})\, d\mathbf{r}, \tag{37}$$

$$\Delta S_{\text{ww}} = -\tfrac{1}{2}k\rho^2 \int g^{(1)}(\mathbf{r})[g^{(1)}(\mathbf{r}')-1][g^{(2)}(\mathbf{r},\mathbf{r}') \ln g^{(2)}(\mathbf{r},\mathbf{r}') - g^{(2)}(\mathbf{r},\mathbf{r}')+1]\, d\mathbf{r}\, d\mathbf{r}', \tag{38}$$

where ρ is the solvent number density, $\rho g^{(1)}(\mathbf{r})$ is the local density of the solvent located at $\mathbf{r}$, $u_{\text{sw}}(\mathbf{r})$ is the interaction potential between the macromolecule and a solvent molecule, and $g^{(2)}(\mathbf{r}, \mathbf{r}')$ and $u_{\text{ww}}(\mathbf{r}, \mathbf{r}')$ are, respectively, the pair correlation function and the interaction potential between two solvent molecules, one at $\mathbf{r}$ and the other at $\mathbf{r}$. Equations (35)–(38) involve certain approximations, in particular, the assumption that correlations involving more than two particles can be neglected; the details are given by Lazaridis (1998a). In eq. (32), ΔV^{slv} is the excess partial molar volume of the solute; that is

$$\Delta V^{\text{slv}} = \int [1 - g^{(1)}(\mathbf{r})]\, d\mathbf{r}. \tag{39}$$

The $P\Delta V^{\text{slv}}$ term can be neglected under ambient conditions. Equations (35)–(38) are written here for monatomic solvent particles, but can be generalized to polyatomic molecules by adding integrations over orientational degrees of freedom (Lazaridis, 2000).

The expression for the solute–solvent energy, E_{sw}, is a straightforward extension of the energy equation of statistical thermodynamics. The solute–solvent entropy, S_{sw}, arises from correlations between the solute and the solvent. These include positional correlations, described by the density oscillations around the solute, and orientational correlations (i.e., the fact that solvent preferentially adopts certain orientations with respect to the solute). The solvent reorganization energy (ΔE_{ww}) and entropy (ΔS_{ww}) account for changes in solvent–solvent interactions and correlations upon solute insertion. All the terms in eqs. (33) and (35) are largest close to the inhomogeneity (i.e. the solute) and decay to zero far from the solute.

The components of the solvation free energy are written as integrals over the space around the solute; that is

$$\Delta G^{slv} = \int f(\mathbf{r}) \, d\mathbf{r}, \tag{40}$$

where $f(\mathbf{r})$ is the solvation free-energy density. Neglecting the $P\Delta V^{slv}$ term, we obtain from eqs. (33) and (34) that

$$f(\mathbf{r}) = \rho g^{(1)}(\mathbf{r}) u_{sw}(\mathbf{r}) + \tfrac{1}{2}\rho^2 g^{(1)}(\mathbf{r}) \int [g^{(1)}(\mathbf{r}') - 1] g^{(2)}(\mathbf{r},\mathbf{r}') u_{ww}(,\mathbf{r}') \, d\mathbf{r}' + \\ kT\rho g^{(1)}(\mathbf{r}) \ln g^{(1)}(\mathbf{r}) + \\ \tfrac{1}{2}k\rho^2 g^{(1)}(\mathbf{r}) \int [g^{(1)}(\mathbf{r}') - 1] g^{(2)}(\mathbf{r},\mathbf{r}') \ln g^{(2)}(\mathbf{r},\mathbf{r}') - g^{(2)}(\mathbf{r},\mathbf{r}') + 1] \, d\mathbf{r}'. \tag{41}$$

This approach has a number of advantages over the traditional solvation theories mentioned above. First, it provides an explicit connection between solvation thermodynamics and solvent structure around the solute, and gives a detailed decomposition of the solvation free energy. It has been used to analyze the thermodynamics of hydrophobic hydration (Lazaridis and Paulaitis, 1992; Lazaridis and Paulaitis, 1994; Lazaridis, 2000) as well as the solvation in simple fluids (Lazaridis, 1998b). This clearly is useful for the physical understanding of the solvation process. Because it gives the solvation free-energy components as integrals over space, it leads to a "modular" approach to the solvation free energy of large molecules. To calculate the difference in solvation free energy between two molecules that differ in one group, one needs only to calculate the integrals in the region around that group. Moreover, this concept can be used to transfer small-molecule information to the study of macromolecular solvation. For example, consider an isolated methyl group and a methyl group in a macromolecule. To the extent that the solvent structure next to the methyl group is the same as that next to an isolated methyl group, the solvation free energy coming from that region of space will be the same. In actuality this is not exact, and so corrections may have to be introduced (Lazaridis and Paulaitis, 1992; Lazaridis and Paulaitis, 1994; Lazaridis, 2000; Lum et al., 1999). Nevertheless, this concept is used in section 6 to develop a simple analytical model for the solvation free energy and other solvation properties of proteins.

5. Empirical Solvation Models

Because the development of solvation models for macromolecules based purely on statistical mechanics has proven difficult, a number of empirical approaches have been proposed. The simplest is the atomic solvation parameter model (Eisenberg and McLachlan, 1986). In this model, the solvation free energy is given as a sum of atomic contributions. The solvation free energy of a group is assumed to be proportional to its accessible surface area A_i:

$$\Delta G^{slv} = \sum_i \sigma_i A_i, \qquad (42)$$

where the proportionality coefficients σ_i depend on the type of atom and are determined by fitting experimental data. We use the Gibbs free-energy symbol as is customary, but ΔG^{slv} is equal for all practical purposes, to the corresponding Helmholtz free energy, as pointed out earlier. There are two types of model that differ in the type of data that are used to determine the parameters. Models which view the protein interior as a nonpolar solvation medium use data for the transfer of molecules from nonpolar liquids to water (Eisenberg and McLachlan, 1986). The second type uses data from the transfer of molecules from the gas phase to water (Ooi et al., 1987; Wesson and Eisenberg, 1992). The latter solvation models can be added to molecular-mechanical force fields. Fraternali & van Gunsteren (1996) used a very simple surface-area-based model for molecular dynamics simulations.

Other empirical solvation models do not use the accessible surface area. For example, the hydration-shell model (Kang et al., 1987) assumes that the hydration free energy of a group arises from the first hydration shell and that it is proportional to the volume of the hydration shell that is accessible to the solvent (i.e., that is not occupied by other solute atoms). Another type of solvation model is based on the contacts each group makes with other solute atoms (Colonna-Cesari and Sander, 1990). The more contacts there are, the smaller is the magnitude of the solvation free energy of the group, and the contacts are weighted according to some function that depends on their distance from the group. This model is much faster to use, because counting the number of contacts takes considerably less time than calculating the surface area. A version of this model was parametrized according to solvation free energies of small molecules (Stouten et al., 1993). The model assumes a linear relationship between the solvation free energy and a weighted sum of the contacts that the group makes with other solute atoms. A solvation parameter is assigned to each group.

These models can be used not only for the solvation free energy but also for the enthalpy and the heat capacity of solvation. Makhatadze & Privalov (Makhatadze and Privalov, 1993) developed such a model assuming that the solvation enthalpy of both polar and nonpolar groups is proportional to the surface area. Evidence for the breakdown of this assumption was obtained by theoretical methods (Lazaridis et al., 1995). The first test was based on the RISM integral-equation theory applied to N-methyl acetamide, the alanine dipeptide, and the alanine tetrapeptide. The solvation enthalpy was calculated for a large number of conformations. It was found that, for nonextended conformations of the alanine tetrapeptide, the CONH group had solvation enthalpies lower than the surface area proportionality assumption would predict. The second test involved the calculation of carbonyl–solvent interactions from a molecular-dynamical simulation of a protein in solution. While this is only one component of the solvation enthalpy (the other is the solvent reorganization energy (Yu and Karplus, 1988)), it gives an indication of the validity of this assumption. The plot of the carbonyl interaction energies against (ASA) exhibited considerable scatter. As with the accessible surface area integral-equation results, it was found that ASA underestimated the interactions of buried groups with the solvent. Even groups with zero ASA can interact significantly with the solvent. This may result in overestimation of the solvation enthalpy change upon denaturation.

Another set of solvation models treats the entire protein at once and is based on continuum electrostatics and the Poisson–Boltzmann (PB) equation (Sharp and Honig, 1990). This method evaluates only the electrostatic component of solvation. The solute is treated as a low-dielectric cavity in a high-dielectric medium. This approach assumes that the electrostatic component of solvation free energy is described adequately by continuum electrostatics, and that the laws of continuum electrostatics hold down to the atomic scale. To obtain a complete solvation model, the nonpolar component of solvation is added—usually a term equal to the accessible surface times a surface-tension-like coefficient. The electrostatic treatment involves the numerical solution of the equation which can be applied to realistic solute geometries (e.g., x-ray or NMR structure of a protein) (Sharp and Honig, 1990; Honig and Nicholls, 1995):

$$\nabla \cdot [\varepsilon(\mathbf{r})\nabla \cdot \phi(\mathbf{r})] - \varepsilon(\mathbf{r})\kappa(\mathbf{r})^2 \sinh[\phi(\mathbf{r})] = -4\pi\rho(\mathbf{r}), \qquad (43)$$

where ε is the dielectric constant, ϕ is the electrostatic potential, ρ is the charge density, and κ is related to the Debye length. Given ε and $\rho(\mathbf{r})$, this equation is solved by numerical methods such as finite differences or finite elements to give the electrostatic potential at every position $\mathbf{r}$. From this, one can calculate the solvation free energy and other useful quantities, such as interactions between particular groups or the effective dielectric constants for the system (Bashford and Karplus, 1990). For a simple spherical solute, the PB equation reduces to the Born model of ion solvation. Applications of the continuum-electrostatics method include calculation of solvation free energies of small molecules (Sitkoff et al., 1994), calculation of pK_a values (Bashford and Karplus, 1990; Yang et al., 1993; Alexov and Gunner, 1997; van Vlijmen et al., 1998), and analysis of the electrostatic contributions to protein and nucleic acid stability (Yang and Honig, 1992).

Although the PB approach is computationally more efficient than free-energy simulations, it still is too computationally demanding to use in a molecular dynamics simulation. Semianalytical or analytical approximations that are much faster to evaluate have been proposed. Still et al. (1990) introduced a simple generalization of the Born formula to polyatomic molecules. More recently, the generalized Born equation was combined with an integrated-field method for self-energies to give a completely analytical treatment of electrostatic energies and forces (Schaefer and Karplus, 1996); successful applications have been made to conformational equilibria of peptides (Schaefer et al., 1998).

6. An Effective-energy Function

In section 4 it was shown that the solvation free energy can be written as an integral over the space around the solute; see eq. (40). This can be used as a basis for developing analytical solvation models. By the use of theory or simulations, one can estimate how much of the solvation free energy comes from the first solvation shell, the second solvation shell, and so on. Such calculations for specific components of the solvation free energy have been performed for a few simple systems (Lazaridis and Paulaitis, 1992; Lazaridis and Paulaitis, 1994; Matubayasi et al., 1994; Lazaridis and Karplus, 1996; Lazaridis, 1998b). These studies have shown that a large portion of the solvation entropy or solvation energy for nonpolar and polar solutes (70 to 90%) arises from the first solvation shell. On the basis of these results, it is reasonable to assume that $f(\mathbf{r})$ for uncharged groups is short-ranged; that is, it decays to zero within the second solvation shell. A simple Gaussian function, with the appropriate correlation length was found to exhibit such behavior, and has been used in the Gaussian exclusion model for solvation thermodynamics (Lazaridis and Karplus, 1999b).

When a solute j approaches solute i, the solvation free energy of i is modified, because j excludes solvent from the volume it occupies. If this was the only effect of j, the solvation free energy of i and j is close would be

$$\Delta G_i^{\text{slv}} = \Delta G_i^{\text{ref}} - \int_{V_j} f_i(\mathbf{r})\, d\mathbf{r}, \qquad (44)$$

where ΔG_i^{ref} is the solvation free energy of isolated i and the integration is over V_j, the volume excluded by j; the variable $\mathbf{r}$ here stands for the position of j relative to i. This is the excluded-volume effect. The second effect is that the structure of the solvent in the region not occupied by j is modified by j, and this may affect the interactions of i with the solvent, especially if i and j are polar (fig. 1.3). This is the *solvent perturbation* effect.

Based on this physical principle, we have developed a model for the solvation free energy and combined it with the CHARMM polar hydrogen energy function to obtain an approximation to the effective energy func-

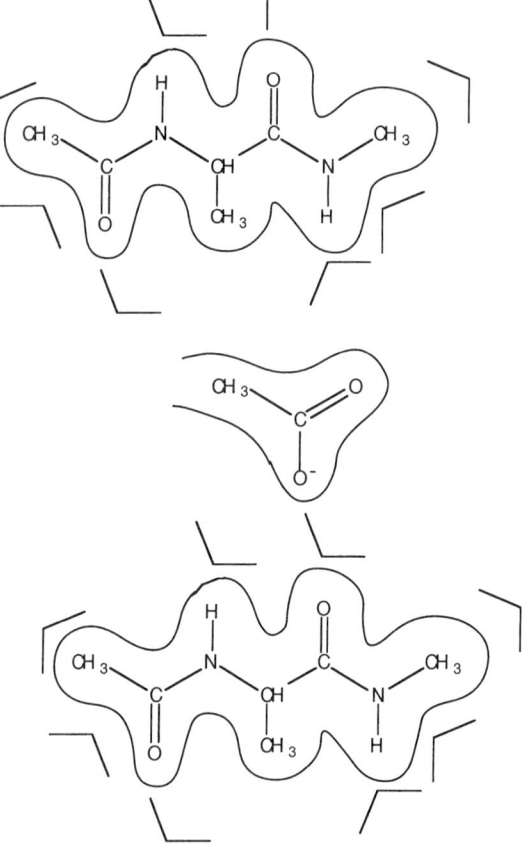

Figure 1.3. Perturbation of solvent structure by neighboring groups

tion (EEF1) (Lazaridis and Karplus, 1999b). The model assumes that, for a polyatomic solute, we can write the solvation free energy as a sum over group contributions (see section 9 and the appendix); that is,

$$\Delta G^{\text{slv}} = \sum_i \Delta G_i^{\text{slv}}, \tag{45}$$

where ΔG_i^{slv} is the solvation free energy of group i. Taking into account only the solvent exclusion effect, we can write

$$\Delta G_i^{\text{slv}} = \Delta G_i^{\text{ref}} - \sum_j \int_{V_j} f_i(\mathbf{r}) \, d\mathbf{r}, \tag{46}$$

where ΔG_i^{ref} (the reference solvation free energy) is the solvation free energy of group i in a suitably chosen small molecule in which group i

is essentially fully exposed to the solvent. The integral in eq. 45 is over the volume V_j of group j and the summation is over all groups j around i. To simplify the calculation the integral over $f_i(\mathbf{r})$ is approximated by the product $f_i(r_{ij})V_j$, where $V_j = \int_{V_j} d\mathbf{r}$, so that

$$\Delta G_i^{\text{slv}} = \Delta G_i^{\text{ref}} - \sum_{j \neq 1} f_i(r_{ij})V_j, \qquad (47)$$

where r_{ij} is the distance between i and j. The solvation free-energy density is assumed to be given by the Gaussian function

$$4\pi r^2 f_i(r) = \alpha_i \exp(-x_i^2), \qquad x_i = \frac{r - R_i}{\lambda_i}, \qquad (48)$$

where R_i is the van der Waals radius of i (half of the distance to the energy minimum in the Lennard–Jones potential), λ_i is a correlation length, and α_i is a proportionality coefficient given by

$$\alpha_i = 2\Delta G_i^{\text{free}}/\sqrt{\pi}\lambda_i \qquad (49)$$

where ΔG_i^{free} is the solvation free energy of the free (isolated) group i; ΔG_i^{free} is close to ΔG_i^{ref}, but not identical to it, and is determined empirically by requiring that the solvation free energy of deeply buried groups be zero.

Equation 46 accounts for the solvent-exclusion effect. The solvent perturbation effect, which is most important for polar and charged groups, is approximately taken into account in the model by using neutralized forms of the ionic sidechains and a linear distance-dependent dielectric constant ($\varepsilon = r$). Both van der Waals and electrostatic interactions are cut off at 0.9 nm (nanometers) with a switching function between 0.7 and 0.9 nm. Electrostatic interactions are calculated on a group by group basis. The value of λ_i was taken to be the thickness of one hydration shell (0.35 nm)—except for the neutralized ionic groups, for which a larger value (0.6 nm) was used. The same model can be employed to calculate other solvation properties, such as the solvation enthalpy, entropy, and heat capacity, by introducing the appropriate reference solvation values for the enthalpy and heat capacity; the solvation entropy is obtained by difference from the solvation free energy and enthalpy.

EEF1 has been tested extensively. It gives stable structures for native proteins during molecular dynamics simulations at room temperature, with modest deviations from the crystal structure (Lazaridis and Karplus, 1999b), it discriminates native from misfolded conformations (Lazaridis and Karplus, 1999a) and gives unfolding pathways at high temperatures in agreement with explicit water simulations (Lazaridis et al., 1997). Nevertheless, the model has a number of deficiencies: it does not account for the directionality of polar-group–solvent interactions, assumes that all empty space is occupied by solvent, and neglects solvent orientational polarization effects. These deficiencies can be rectified by refinements (work in progress), inevitably at the expense of simplicity and computational efficiency.

Unlike many of the knowledge-based energy functions, which appear to be limited to the evaluation of protein conformations obtained by threading and related procedures, EEF1 is applicable to protein folding and unfolding studies by molecular dynamics simulations, for example. The formulation proposed is only 50% slower than a vacuum simulation, and thus makes possible many studies for which simulations in explicit water are prohibitively expensive. Given its physical basis and decomposability, the effective energy function can be used for approximate thermodynamic analysis of contributions to protein stability.

7. Contribution of Intramolecular Interactions to Protein Stability

The intramolecular energy term in eq. (29) includes contributions from bonded terms (bond stretching and bending, torsional potentials, deviations from planarity of aromatic rings, etc.), and nonbonded terms (dispersion and electrostatic interactions, including hydrogen bonding). The bonded terms are not expected to be substantially different in the folded and unfolded conformations. The largest contribution to $\Delta \langle H_{mm} \rangle$ is expected to come from the nonbonded interactions. Nonbonded interactions, of course, exist between the solute and the solvent, but in the present analysis these are included within the solvation free-energy term.

A schematic construct that can be used to visualize the partitioning of the free energy in eq. (29) is the thermodynamic cycle of fig. 1.4 (Lazaridis et al., 1995). The unfolding reaction is imagined to occur in the gas phase in the same way as it occurs in solution; that is, the same conformational ensembles correspond to the native and the unfolded state. The terms in eq. (29) relevant to the gas-phase reaction are $\Delta \langle H_{mm} \rangle$ and $T \Delta S^{conf}$. The vertical processes correspond to inserting the folded and unfolded ensembles into the solvent. They give rise to the ΔG^{slv} terms for the native (N) and unfolded (U) configuration states. The total ΔG in eq. (29) is obtained by completing the thermodynamic cycle in fig. 1.4. The same cycle can be used to analyze the enthalpy of unfolding. In that case, $\Delta \langle H_{mm} \rangle$ is the relevant quantity for the gas phase and $\Delta \Delta H^{slv}$ for the vertical reactions. The enthalpy of unfolding in solution is

$$\Delta H = \Delta \langle H_{mm} \rangle + \Delta \langle \Delta H^{slv} \rangle. \tag{50}$$

Several functions (section 3) that describe the intramolecular energy of the protein are available (e.g., Brooks et al., 1983; Weiner et al., 1986) and can be used, in principle, to evaluate the intramolecular contribution to the enthalpy and free energy of unfolding. Since the native state energy can be evaluated by molecular dynamics simulations starting with the known structure, it is necessary only to construct a model for the con-

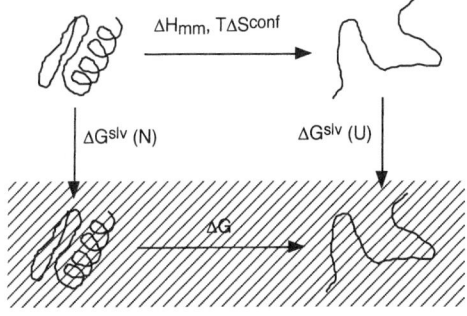

Figure 1.4. Thermodynamic cycle for protein folding

formational ensemble corresponding to the denatured state. In previous work, it has been customary to approximate the properties of the unfolded state by summing the properties of individual amino acid residues. A better model, though still approximate, is to use a fully extended polypeptide chain to represent the unfolded state.

Using the latter model for the unfolded state, the $\Delta \langle H_{mm} \rangle$ term was calculated with the CHARMM polar hydrogen energy function and neutralized sidechains (Lazaridis et al., 1995). Table 1.1 shows the results of this calculation as well as the decomposition of this term into van der Waals, electrostatic, and bonded contributions. The calculated values for $\Delta \langle H_{mm} \rangle$ ranged from 943 kcal/mol for cytochrome c to 1492 kcal/mol for myoglobin. It was indeed found that the bonded terms make a very small contribution to $\Delta \langle H_{mm} \rangle$. The quantity $\Delta \langle H_{mm} \rangle$ was shown to consist 60–70% of van der Waals interactions and 25–35% of electrostatic interactions; hydrogen bonding arises mainly from the electrostatic term in the energy function (eq. (31)).

The van der Waals term was further decomposed into contributions from nonpolar–nonpolar, polar–polar, and polar–nonpolar interactions

Table 1.1. Calculated Vacuum Enthalpy of Unfolding

Protein	N_{res}	MW	ΔH_N^U (vac)	ΔH_N^U (vdW)	ΔH_N^U (elec)	ΔH_N^U (bond)
Cytochrome C	103	12300	943	688	225	30
RNAse A	124	13700	1068	654	398	16
Lysozyme	129	14300	1116	738	351	27
Myoglobin	153	17800	1492	1020	421	51

All values in kcal/mol calculated as described in the two paragraphs following eq. (50); see also the appendix and table V of Lazaridis et al. (1995).

For myoglobin and cytochrome C, it is assumed that the free heme will have the same self (intraheme) energy as in the protein.

Table 1.2. Van der Waals contributions to the unfolding enthalpy, $\Delta H_N^U(\text{vac})$

Protein	$\Delta H_N^U(\text{vdW})$			
	np–np	np–p	p–p	Total
Cytochrome C	211	372	113	688
RNAse A	148	366	140	654
Lysozyme	205	396	137	738
Myoglobin	306	573	144	1020

All values in kcal/mol; see table VII of Lazaridis et al., (1995).

(table 1.2). Nonpolar atoms are all carbons except the backbone carbonyl carbon and the polar carbon atoms of the Asp, Glu, Asn, and Gln sidechains. All other atoms are considered to be polar. What is perhaps surprising in table 1.2 is the large contribution from polar–nonpolar interactions. This shows that the interactions in the protein interior are rather complex and cannot be represented as a simple sum of hydrophobic interactions and hydrogen bonds.

The estimates for the change in intramolecular enthalpy upon folding can be combined with solvation enthalpies calculated from the accessible surface area (Makhatadze and Privalov, 1993) to obtain estimates of the enthalpy of denaturation. The results of the calculation for four proteins are shown in table 1.3 (Lazaridis et al., 1995). The values for ΔH are large and negative, whereas the experimental values are small and positive. Part of the discrepancy arises from the assumption of the proportionality of solvation enthalpy to the accessible surface area, particularly for the polar terms (see section 5). If one assumes that the major error arises from the polar term, the reduced values in parentheses are required to obtain agreement with experiment; see also Table E-IV of Lazaridis et al. (1995). More important is the fact that the fully extended model used for the denatured state is unrealistic. These issues are further explored in the following section.

The Poisson–Boltzmann approach has also been used to estimate the electrostatic solvation free energy for the native and unfolded forms of

Table 1.3 Estimate of the Enthalpy of Denaturation

Protein	ΔH_N^U (vac)	ΔH_N^U (sol, np)	$\Delta H_N^U(\text{sol, p})$	ΔH (calc)	ΔH (exp)
Cytochrome C	943	−172	−1067 (−750)	−296	21
RNAse A	1068	−159	−1127 (−838)	−218	71
Lysozyme	1116	−192	−1224 (−866)	−300	58
Myoglobin	1492	−176	−1541 (−1214)	−325	1.4

All values in kcal/mol; see also table 1.1

Table 1.4 Electrostatic Solvation (Free) Energy Differences of Folded and Unfolded Proteins in the Poisson–Boltzmann Approximation[a]

Protein	$\Delta G_{vac}^{sol}(N)$	$\Delta G_{vac}^{sol}(U)$	$\Delta G_N^U(sol,p)$	ΔH_N^U (vac,elec)	ΔG_N^U (sol,elec)[b]
Cytochrome C	−441	−789	−348	225	−123
RNAse A	−539	−958	−419	398	−21
Lysozyme	−546	−1020	−474	351	−123
Myoglobin	−558	−1087	−529	421	−108
ALA20	−45	−89	−44	60	16

[a] All values in kcal/mol
[b] $\Delta G_N^U(sol,elec) = \Delta G_N^U(sol,p) + \Delta H_N^U(vac,elec)$

these four proteins and for a 20-residue polyalaline helix (Lazaridis et al., 1995). The results are shown in table 4. The first two columns of entries in table 1.4 are the Poisson–Boltzmann electrostatic solvation free energy of the native and unfolded protein. The third column is the difference of the two—that is, the change in solvation free energy upon unfolding. The electrostatic solvation energies are more negative for the unfolded conformations, primarily because backbone hydrogen-bonding groups that are buried in the native state become exposed to the solvent. The fourth column is the change in intramolecular electrostatic enthalpy upon unfolding. The last column is the sum of the third and fourth columns and gives the estimated contribution of electrostatic interactions to unfolding in solution. The latter is found to be negative for the four proteins (i.e., it stabilizes the unfolded state) and positive for the polyalanine helix.

If we use the results for the electrostatic solvation free energy as an estimate of the change in solvation enthalpy to close the cycle in fig. 1.4, the resulting unfolding enthalpies would be too positive. This is probably due to the fact that these values correspond to solvation free energies, not enthalpies, and include a positive entropic contribution (the solvation entropy is negative). Also, the nonelectrostatic component of the solvation enthalpy is not included in these values. Finally, the problem with the fully extended model of the unfolded state is still present.

The intramolecular energy terms and their decomposition into polar and nonpolar contributions together with the empirical estimates of solvation enthalpy and entropy contributions can be used to address the question of the relative contribution of polar and nonpolar groups to protein stability. The contribution of the nonpolar groups to the enthalpy is defined as (a) the nonpolar–nonpolar van der Waals term, (b) one half of the nonpolar–polar van der Waals term, and (c) the nonpolar solvation enthalpy change upon unfolding. The contribution

Table 1.5. Proposed Contributions of Polar and Nonpolar Groups to $\Delta H_N^U(\text{sol})$ at 25 °C

Protein	Polar	Nonpolar
Cytochrome C	−204	225
RNAse A	−101	171
Lysozyme	−152	210
Myoglobin	−315	316

of the polar groups is equal to (a) the polar–polar van der Waals term, (b) the electrostatic energy term, (c) one half of the polar–nonpolar van der Waals term, and (d) the polar solvation enthalpy change upon unfolding. The results are shown in table 1.5. The polar groups are seen to make a negative contribution and the nonpolar groups a positive contribution to the enthalpy of unfolding. If we add the solvation entropy estimates (see table 1.6), then the nonpolar groups are seen to make the dominant contribution to the free energy of unfolding, while the polar groups make a smaller or zero contribution. This is consistent with the traditional idea (Kauzmann, 1959) that the hydrophobic interaction provides the major driving force for folding. The role of the polar groups is to make the protein soluble in water and to introduce specificity in the low-free-energy conformations that make up the native protein; that is, the many compact structures that exist are not competitive in free energy with the native structure due to poor polar interactions.

8. The Denatured State of Proteins

In past work, the model adopted for the denatured state was either the sum of individual amino acids (Miller et al., 1987; Makhatadze and Privalov, 1993) or an extended, completely solvent-exposed, polypep-

Table 1.6 Contribution of Polar and Nonpolar Groups to the Entropy, $-T\Delta S_N^U(\text{sol})$, and Free Energy, $\Delta G_N^U(\text{sol})$, of Protein Unfolding[a]

| | $-T\Delta S_N^U(\text{sol})$[b] | | $\Delta G_N^U(\text{sol})$ | |
Protein	Polar	Nonpolar	Polar	Nonpolar
Cytochrome C	202	208	−2	433
RNAse A	225	200	124	371
Lysozyme	239	240	87	450
Myoglobin	289	340	−26	656

[a]All values in kcal/mol at 25 °C; see table E-VI of Lazaridis et al. (1995).
[b]Values from Makhatadze and Privalov (1995).

tide chain (Makhatadze and Privalov, 1994; Lazaridis, et al., 1995; Makhatadze and Privalov, 1995). This assumption is at odds with experimental evidence showing that the denatured state in the absence of denaturants is rather compact (Dill and Shortle, 1991; Shortle, 1996). Residual structure has been found in the heat-denatured states of many proteins (Ptitsyn, 1995; Nölting et al., 1997). Theoretical work on simplified protein models (Shortle et al., 1992; Lattman et al., 1994) and molecular dynamics simulations in explicit solvent (Bond et al., 1997; Kazmirsky and Daggett, 1998) also suggest a relatively compact denatured state. Residual structure and compactness implies that a significant fraction of the protein residues are interacting with each other and are, at least in part, sequestered from solvent, although the fluctuations in the structures are such that little hydrogen-exchange protection is present.

The model of a completely unfolded denatured state has arisen in part from calorimetric measurements of the heat capacity change of protein denaturation. It is generally believed that the positive ΔC_p values for unfolding arise primarily from the exposure of nonpolar groups to water (Brandts, 1964; Sturtevant, 1977; Baldwin, 1986; Spolar et al., 1989; Makhatadze and Privalov, 1990; Privalov and Makhatadze, 1992). The basis for this conclusion is that the transfer of nonpolar groups from the gas phase or nonpolar liquids into water is also accompanied by a large positive heat-capacity change (Privalov and Gill, 1988). More recently, it was realized that polar groups also make a contribution to ΔC_p that is smaller than, and of the opposite sign to, that for nonpolar groups (Makhatadze and Privalov, 1990; Murphy and Gill, 1991; Spolar et al., 1992). It has been found that ΔC_p can be well reproduced by assuming a fully unfolded denatured state (all residues fully exposed to solvent (Privalov et al., 1989)) if ΔC_p is taken to be proportional to the change in the exposed surface area on unfolding. In such calculations, the proportionality constants appropriate for the amino acids are obtained from transfer experiments with small model compounds (Makhatadze and Privalov, 1990; Murphy and Freire, 1992; Spolar et al., 1992; Gomez et al., 1995). Thus, there is an apparent discrepancy between the realistic description of the denatured state as being rather compact and the semiempirical ΔC_p models which yield good agreement with experiment if a fully extended denatured state is used.

To resolve this discrepancy, we have used EEF1 to generate models for the denatured state by molecular dynamics simulations (Lazaridis and Karplus, 1999c). The system chosen for this analysis was the 64-residue truncated version of the small protein CI2, for which experimental data and unfolding simulations are available. The heat capacity was calculated by performing simulations of the native (N) and denatured (D) conformations at 280 K and 320 K, calculating the enthalpy at these temperatures using the solvation enthalpy parameters in EEF1, and taking the finite difference:

$$\Delta C_{\mathrm{p}} = \left(\frac{\partial \Delta H}{\partial T}\right)_{\mathrm{P}} = \frac{\Delta H(320\,\mathrm{K}) - \Delta H(280\,\mathrm{K})}{40}. \tag{51}$$

Although the absolute partial molar heat capacity of a protein cannot be calculated quantitatively by classical mechanics (Brooks and Karplus, 1983), quantum effects cancel out approximately when the difference in heat capacity between two protein states is taken. Since the enthalpy can be decomposed into intramolecular and solvation terms, the heat capacity can likewise be decomposed. The calculation of the heat capacity takes into account both the intrinsic temperature dependence of the solvation enthalpy of the protein groups and the contribution from the temperature dependence of the protein's conformation distribution $p(\mathbf{q})$ (see section 4). The latter contribution has been neglected in empirical models that use the fully extended chain as a model of the denatured state. An additional small contribution—that from redistribution of the protein population among different local minima in the denatured state—is neglected because it would require extensive sampling of the conformational space.

Three denatured state models were generated by molecular dynamics simulations starting from a fully extended chain. They all resulted in relatively compact structures. The effective energies of the three denatured states were between 38 and 65 kcal/mol higher than that of the native state. This is a reasonable difference, considering the experimental protein stability and the estimated change in conformational entropy (Lazaridis and Karplus, 1999b). The radius of gyration (R_g) of the denatured conformations was only 12–18% greater than that of the N state, in agreement with experimental data. The "hydrophobic collapse" from the extended state found in the molecular dynamics simulations was accompanied by formation of a large number of protein–protein hydrogen bonds; the number of hydrogen bonds in the denatured forms was only slightly smaller than that in the native state, but very few were native-like. Moreover, the three denatured conformations had almost no native contacts (pairs of atoms more than three residues apart in sequence that are within 0.4 nm from each other in the crystal structure (Lazaridis and Karplus, 1997)).

It was found that the average value for ΔC_{p} of unfolding is 0.525 ± 0.15 kcal/mol K, compared to the experimental value of 0.789 kcal/mol K (Jackson and Fersht, 1991); the experimental value pertains to the untruncated form of CI2, but the contribution of the disordered region to ΔC_{p} is expected to be small. The underestimation of ΔC_{p} may be due to deficiencies of the hydration model (for example the group additivity assumption), to the obviously limited sampling of the denatured state or to the neglect of transitions on a longer timescale (see above). Covalent contributions are approximately equal in the denatured and native conformations, and thus do not make a significant contribution to the unfolding heat capacity, as expected (Karplus et al., 1987; Gomez et al., 1995). In one of the three denatured models, the solvation contribution was found to be negative, despite the fact that exposure of nonpolar groups is higher in the denatured state. This was due to the fact that the rise in temperature increased the exposure of polar groups in the denatured state and thus decreased the solvation enthalpy difference between the denatured and native conformation. For one of the denatured conformations, the intramolecular contribution was very large (0.95 kcal/mol K); for the other two it was smaller (0.05 and 0.125 kcal/mol K), but still significant.

The major conclusion from this simulation study is that a significant contribution to the heat capacity of denaturation comes from protein–protein nonbonded interactions and their changing contribution as a function of temperature. The denatured state D is relatively compact but more labile than the native state so that temperature can break interactions in D more easily than in N. This means that the enthalpy of D increases with temperature more than that of N and contributes significantly to ΔC_p. Moreover, the results demonstrate that relatively compact denatured states are qualitatively consistent with the calorimetric data for ΔC_p. The smaller contribution of hydrophobic-group exposure to the heat capacity in this model for the denatured state is compensated by the contribution of noncovalent protein interacations.

9. Defining Contributions to Stability

Analyses of protein stability attempt to provide an understanding of the contributions of different types of interaction. This is exactly what was done in the previous section, where the unfolding enthalpy was decomposed into (a) intraprotein-bonding van der Waals and electrostatic terms and (b) solvation terms. If one wishes to decompose the intraprotein van der Waals terms further into polar and nonpolar contributions, as is often done, the large polar–nonpolar van der Waals interactions have to be divided between the two; see table XXII of Lazaridis et al. (1995) and the associated discussion. This is a simple example of the fact that some assumptions are required in decomposing the enthalpy, and also the free energy, into contributions that are useful for understanding and for predictions. The complexity of the problem is such that there has been considerable discussion recently concerning the validity of such decompositions, as described below. It has become clear that the decompositions are meaningful, though it is necessary to interpret them carefully and understand the assumptions involved. Even the authors who most strongly criticized the decompositions of the free energy have begun to claim them as their own—in studying, for example, ligand–protein interactions by perturbation models (Gerber et al., 1993).

Two types of decomposition are in common use. The first is a decomposition of the total free energy of folding according to the type of interaction; the second is a decomposition of the free energy into contributions from constituent groups. The latter is particularly useful for the interpretation of site-directed mutagenesis experiments, where one residue is replaced by another, and where the effect of the substitution on protein stability or the value of a ligand binding constant is determined. Such decompositions have often been performed on a somewhat intuitive basis, although careful discussions of the assumptions involved have been given in some of the experimental analyses (Fersht, 1988). As mentioned above, it has been argued that such decompositions are not possible, because entropy is a global property and cannot be dissected (Mark and van Gunsteren, 1994; Dill, 1997). This conclusion is uninformed, as we show by providing a theoretical basis for such decompositions and by presenting a clear definition of group contributions to stability.

Thermodynamic integration offers an exact way of decomposing a free-energy difference into contributions from interactions or groups (Gao et al., 1989). The basic formula for thermodynamic integration is

$$\Delta A = \int_0^1 \left\langle \frac{\partial U(\lambda)}{\partial \lambda} \right\rangle_\lambda d\lambda, \tag{52}$$

where $U(\lambda)$ is an empirical energy function (e.g., eq. (31)) that describes the initial state for $\lambda = 0$ (e.g., the wild-type protein) and the final state for $\lambda = 1$ (e.g., the mutant protein); that is,

$$U(\lambda) = (1 - \lambda)U_{\text{wild-type}} + \lambda U_{\text{mutant}} \tag{53}$$

The subscript λ on the brackets in eq. (52) indicates that the simulation is done with $U(\lambda)$. Calculations with eq. (52) have been referred to as computer alchemy (Gao et al., 1989; Straatsma and McCammon, 1992), because they involve transforming one molecule into another on the computer. It would be easy to satisfy the alchemist's dream of transforming lead into gold by this technique.

Since $U(\lambda)$ can be decomposed into contributions from various types of interaction, and since eq. (52) is linear in $U(\lambda)$, the free energy itself can be decomposed in a corresponding fashion. To obtain the effect of a point mutation on protein stability, for example, the protein is simulated in both the native and denatured states starting with the wild-type sequence and reversibly changing the energy function to that of the mutant (see fig. 1.5). It has been realized that this type of decomposition is dependent on the integration path taken in the transformation of one group into another (Brooks, 1990; Simonson and Brünger, 1992). Despite the path dependence, these decompositions are useful (Boresch et al., 1994; Boresch and Karplus, 1995). Formal analyses of these decompositions have been presented (Brady and Sharp, 1995; Archontis and Karplus, 1996; Brady et al. 1996).

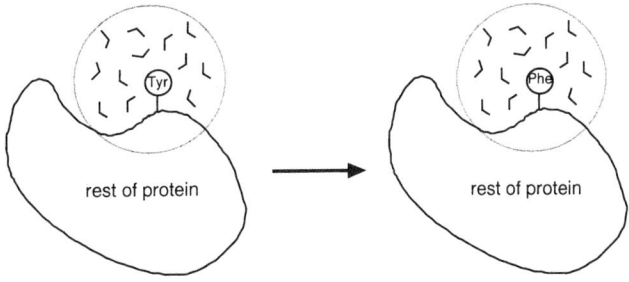

Figure 1.5. Free-energy simulation transforming the wild-type protein with Tyr in the binding site to a mutant with Phe in the binding site. The dashed line indicates the spherical region that is studied in the free-energy simulations and the (∧) indicates water molecules solvating the binding site.

One example of such a decomposition that offered useful insights is the analysis of the binding of Tyrosine to tyrosyl-tRNA synthetase, where the difference in free energy of binding of the substrate to the wild type and the Tyr169 → Phe mutant of the enzyme was calculated by thermodynamic integration following eq. (52) (Lau and Karplus, 1994). The experimental analysis of a series of tyrosyl-tRNA mutants, including the Tyr 169 → Phe mutant is given by Wells & Fersht (1986). (An excellent overview of "protein engineering" methods, as such mutation studies are now called, is given in chapter 15 of Fersht (1999).) The calculations showed that the mutation reduces the binding affinity by about 3 kcal/mol—a result in good agreement with experiment. The path chosen in the study was the linear transformation of the hydroxyl group of Tyr 169 to a hydrogen atom in the bound system (enzyme plus substrate) and the free system (see fig. 1.6). From the diagram, the difference $\Delta\Delta G$ in binding free energy between the mutant (ΔG_m) and the wild type (ΔG_w) is equal by Hess's Law (Tembe and McCammon, 1984) to

$$\Delta\Delta G = \Delta G_b - \Delta G_f = \Delta G_m - \Delta G_w, \tag{54}$$

where ΔG_b is the free energy difference between the mutant and the wild type in the bound form, and ΔG_f is the corresponding free energy difference for the free enzyme. The experiments are done following the vertical paths for the wild type and the mutant, while the calculations are done most easily along the horizontal (alchemical) path. It would, of course, be possible to follow the vertical paths, which would provide the potential-of-mean-force curve for binding the tyrosine ligand to the wild-type and mutant protein.

The free-energy difference was decomposed in several different ways. First, it was decomposed into electrostatic, van der Waals, and covalent interactions. The dominant contribution was found to arise from the elec-

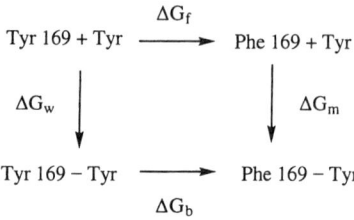

Figure 1.6. Thermodynamic cycle used in the mutation Tyr 196 to Phe

trostatic term. The free-energy difference was also decomposed into interactions of the mutated group with the protein, the ligand, and the solvent (see table 1.7). As expected, the largest contributions were made by the interactions with the ligand and the solvent. The protein contribution was further analyzed into contributions from the neighboring amino acids. It was found that, although the total protein contribution was small, it contained large compensating contributions from individual amino acids. This is probably due to the fact that the position and orientation of the mutated hydroxyl shifts between the free and bound states, forming favorable interactions with some residues in one case and with other residues in the other. The solvent term was also broken down into contributions from water molecules at different distances. It was found that significant contributions arise from relatively distant water molecules.

Although the analysis just described for decomposing the free energy is very useful, one can ask whether it is possible to define contributions to stability that are path-independent. To better understand the type of decomposition we are interested in, consider a hydrogen bond in the interior of a protein. How could one define precisely the "contribution" of this hydrogen bond to protein stability? Obviously, it is not the interaction energy between the CO and NH groups primarily because the direct CO–NH interaction is only a small part of the interactions in which these two groups are involved in the protein interior (Lazaridis et al., 1995). Moreover, taking into account this direct interaction neglects the fact that these two groups in the unfolded state interact favorably with the solvent. Then the free-energy contribution to stability is determined

Table 1.7 Free-energy Decomposition of Noncovalent Interactions in Tyr→Phe Simulations (kcal/mol)

Term	ΔG_b	ΔG_f	$\Delta\Delta G$
Protein	5.3	6.1	−0.83
Solvent	0.3	5.6	−5.2
Ligand	8.6	–	8.6
Total	14.2	11.7	2.6

both by the folded and unfolded state, as was already pointed out for the free-energy simulations above. Clearly, any definition of the "contribution" of protein groups to stability must consider: (a) the multitude of interactions made by these protein groups and (b) the effects of changes in the solvation of these groups.

The starting point for the present discussion is eq. (29). The decomposition presented in eq. (29) is rigorous as long as the intramolecular interaction terms are separable from the solute–solvent and solvent–solvent interaction terms. This is the case for most energy functions in use today (e.g., eq. (31). Actually, Eq. (29) can be generalized to take account of the presence of three-body forces. For the intramolecular energy of the macromolecule, H_{mm}, we assume that the empirical energy function is valid (see section 3). The nonbonded interactions can naturally, though arbitrarily, be assigned to the interacting atoms: half of the interaction to each partner. Usually, the bonded terms ("strain energy") are treated separately. However, they too can be assigned to the atoms that are involved: two atoms for bonds, three for angles, and four for dihedral angles. In this model, each atom in the macromolecule is assigned one half of its nonbonded binding energy with all other atoms, one half of the covalent-bond energy of the bonds, one third of the bond-angle energy, and one quarter of the dihedral energy in which it participates.

Decomposition of the solvation free energy is more complicated. Equation (10) for the solvation free energy is not very convenient for application to a macromolecule, because it does not easily lend itself to a decomposition of the solvation free energy into contributions from groups. The inhomogeneous theory of Section 4 is more convenient for this purpose. The appendix presents a formal approach for defining group contributions to the solvation free energy. Once group solvation free energies are defined, one can define the effective free energy (GFE) of group i as

$$W_i = H_i + \tfrac{1}{2} \sum_j H_{ij} + \Delta G_i^{slv}, \tag{55}$$

where H_i corresponds to the internal group energy (including the bonded terms within group i and any intragroup nonbonded interactions), H_{ij} corresponds to the interactions between groups i and j, and ΔG_i^{slv} is the solvation free energy of group i (see the appendix and section 4). Based on these results, the free energy of unfolding can be written

$$\Delta G = \sum_j \langle \Delta W_i \rangle - T \Delta S^{conf}, \tag{56}$$

where $\langle \Delta W_i \rangle$ is equal to the contribution of group i to stability (the angle brackets indicate a configurational average):

$$\langle \Delta W_i \rangle = \tfrac{1}{2} \sum_j \langle \Delta H_{ij} \rangle + \langle \Delta H_i \rangle + \langle \Delta \Delta G_i^{slv} \rangle, \tag{57}$$

where the extra deltas in eq. (57), relative to eq. (55), correspond to the difference between the folded and unfolded state. Although $\langle \Delta W_i \rangle$ includes the group's share of the bonded energy $\langle H_i^{bond} \rangle$, it is not expected to make a large contribution when differences between GFEs are considered. As is indicated in eq. (57), $\langle \Delta W_i \rangle$ does not include any conformational entropy. The sum of all group contributions is not the free energy of unfolding, ΔG, but the change in effective energy upon unfolding, ΔW.

This clear definition of group contributions can be of use in interpreting mutation experiments. Among site-directed mutagenesis experiments, the easiest to interpret are the so called "nondisruptive deletions" (Fersht, 1987), where an amino-acid sidechain is truncated to eliminate a particular interaction without disrupting the protein structure or introducing additional interactions. For example, the mutation of an Ile to a Val should give information about van der Waals and hydrophobic interactions, whereas the mutation of a Ser to an Ala should give information about hydrogen-bonding interactions. However, it is important to note (Fersht, 1988, 1999) that elimination of a group by mutation does not measure the contribution of that group to protein stability. Deletion of a group leads to loss of the contribution of the deleted group, and modification of the contributions of the neighboring groups:

$$\Delta\Delta G(i) = \langle \Delta W_i \rangle + \sum_{j \neq i} \Delta \langle \Delta W_j \rangle \qquad (58)$$

Thus, in protein engineering, the sum of the changes in stability or binding affinity upon mutation of a series of groups does not give the total stability or binding affinity, even if the protein structure ($p(\mathbf{q})$) is not affected. This is true because the modification of the contributions of interacting groups is counted twice (Boresch et al., 1994).

In several protein engineering studies, a hydrophobic sidechain, partly or fully buried, was truncated to directly estimate the contribution of hydrophobic interactions to protein stability (Matsumura et al., 1988; Kellis et al., 1989; Shortle et al., 1990; Sandberg and Terwilliger, 1991; Eriksson et al., 1992; Serrano et al., 1992). Simulation studies of these mutations were also undertaken (Prevost et al., 1991; Sneddon and Tobias, 1992). Perhaps the most surprising result was that the magnitude of the destabilization observed was higher than that expected from water-to-octanol transfer free energies of model compounds. Crystallographic studies showed that the magnitude of the destabilization correlated with the size of the cavity left in the mutant when the group was deleted (Eriksson et al, 1992). This finding suggested that the extra destabilization free energy can be attributed to the free-energy cost of having a cavity in the protein interior (Eriksson et al., 1992; Lee, 1993; Rashin, 1993). GFEs offer a particularly simple way of analyzing the problem. As mentioned above, deletion of a group will eliminate its contribution but also modify the GFEs of its neighboring groups. Consider, for example, the Ile→Val mutation, in which one methylene group is removed. In the unfolded state, where the mutated residue is most likely to be mainly exposed to solvent, the deletion eliminates the GFE of a methylene group, which is dominated by the solvation free energy of that group, ΔG_i^{solv}. The GFEs of the surrounding groups may change somewhat, because the elimination of the Cδ group increases the solvent exposure of the Cγ group. In the folded state, the sidechain is buried in the cases studied, so that the GFE of the deleted group will be approximately equal to half of the van der Waals interaction of the group with its surroundings, E_{vdW}. When the group is deleted, assuming that the structure remains constant, what is lost is the whole E_{vdW}, only half of which "belongs" to the deleted group. The rest belongs to the GFEs of the surrounding groups. If the protein relaxes to completely fill the gap, the van der Waals energies of the surrounding groups will approximately return to their original value. Thus, the measured difference in protein stability between the mutant and the wild type corresponds to the intrinsic contribution of the deleted group only in the limit of complete protein relaxation. The van der Waals interaction energy for a methylene group in the protein interior is about -3.5 kcal/mol, compared with about -2 kcal/mol in liquid-like environments (Lazaridis et al., 1995). Therefore, the maximum extra destabilization one can expect for the deletion of a methylene group is half of that, or about 1.75 kcal/mol. The experimental values for Ile → Val mutations are in the range 0.3 to 1.8 kcal/mol (Serrano et al., 1992).

Another area where analysis on the thermodynamics of an atomic basis is of interest concerns the magnitude of contribution of hydrogen bonds to protein stability (Honig and Yang, 1995; Lazaridis et al., 1995; Myers and Pace, 1996). The classical view has been that, upon folding, hydrogen bonds betwen protein atoms replace similar hydrogen bonds with the solvent in the unfolded state, and thus lead to little net stabilization of the native structure. Their major role is then presumed to be that they restrict the number of possible folded structures and thus contribute to the stabilization of a unique structure. Mutagenesis experiments have been interpreted as indicating that hydrogen bonds make a favorable contribution to stability or binding (Fersht, 1987; Shirley et al., 1992). This also emerged in an empirical treatment of protein thermodynamics based on model compound data (Makhatadze and Privalov, 1993, 1994). To analyze these results in the present framework, we have to define the term "contribution of a hydrogen bond" in the context of protein stability. This contribution corresponds to the free-energy gain from the interaction of the hydrogen-bonding groups in the protein minus the cost of desolvating these groups (i.e., having them interacting with solvent in the denatured state). One problem that arises in evaluating this quantity is that it is not well defined, because polar groups in the interior of proteins are involved in a multitude of interactions with other polar groups, some of which are hydrogen-bonding and some of which are not (Lazaridis et al.,

1995). In addition, hydrogen-bonding groups have significant van der Waals interactions with nearby nonpolar groups. The cost of desolvating the two hydrogen-bonding groups cannot be assigned to their pairwise interaction alone, but includes all the interactions in which these groups are involved. Consequently, there is a conceptual difficulty in isolating the "hydrogen-bonding" energy (Lazaridis et al., 1995). For these reasons, it is preferable to speak of contribution of groups (GFEs), rather than of interactions, which, as shown above, are well defined quantities. Calculation of the GFE of a hydrogen-bonding group in the folded and unfolded protein would give the contribution of this group to stability. If the group is buried, then its stability contribution will be favorable if half of the sum of all of its interactions in the protein interior is more negative than its solvation free energy in the unfolded protein. That will be the case, for example, if the group forms multiple hydrogen bonds and possibly other favorable dipole–dipole interactions (Lazaridis et al., 1995) in the protein interior.

10. Future Directions

Equilibrium thermodynamics is an established tool for the study of biological processes in vitro. Knowing the equilibrium properties of macromolecules is an essential first step for the characterization of the processes in which they participate. In combination with molecular models and statistical mechanics, it promises to provide a microscopic understanding of the mechanisms of many of the phenomena operating in living systems.

The cornerstone of a microscopic analysis of macromolecular thermodynamics is the energy function. The size and complexity of biological molecules necessitates the use of approximate energy functions. Although molecular mechanics force fields have deficiencies, the source of the largest errors at this time is in the calculation of the solvation term. Significant efforts are being made in developing quantitative models for the solvation free energy and decomposing them into the enthalpy and the entropy of solvation. Modeling of the solvent should not be restricted to pure water, but should be extended to more complex solvent media, at nonneutral pH that contain ions, cosolvents, cosolutes, and so on, to be able to understand the behavior in the complex cellular medium.

We have used protein folding as the example to illustrate the microscopic analysis of macroscopic thermodynamic data. This is, of course, only one area, albeit an important one, where thermodynamics can be used to study elementary biological process. Another area that is justifiably receiving considerable attention is the process of binding between macromolecules or between a macromolecule and a ligand. Many more complex events occur in the living cell; they include protein translocation through membranes, large conformational changes, active diffusion, DNA unwinding, vesicle budding, membrane fusion, and virus assembly. On an even larger scale, there are processes like cell division, cell differentiation, development, growth, and aging. Thermodynamics provides an approach for understanding the driving forces responsible for all these processes and for rationalizing the observed pathways. As more detailed data become available concerning the specific molecules involved, it will become possible to utilize techniques like the ones described here to achieve a description of these events at the atomic level. Such an understanding may aid in learning how to control these cellular events so as to be able to alter them in desirable directions. This offers hope of aiding in the development of methods leading to the demise of cancerous cells or to an increase in the resistance to viral infections.

Appendix Decomposition of the Solvation Free Energy into Group Contributions

Neglecting the *PV* term, the solvation free energy is the sum of the solvation energy and the solvation entropy. Each of these contains a solute–solvent and a solvent-reorganization term. The contribution that is straightforward to decompose is the solute–solvent energy. In most empirical energy functions the solute–solvent interaction potential is broken up into solute-atom–solvent-atom terms. These can be summed to give

$$u_{sw}(\mathbf{r}) = \sum_i u_{iw}(\mathbf{r}), \tag{A1}$$

where *i* enumerates solute atoms or groups, such as carbonyl carbon or a methyl group, and w refers to all the solvent molecules. Given eq. (A1), the decomposition of the solute–solvent energy, E_{sw}, is straightforward; that is,

$$E_{sw} = \sum_i \int g^{(1)}(\mathbf{r}) u_{iw}(\mathbf{r}) \, d\mathbf{r}, \tag{A2}$$

where $\rho g^{(1)}(\mathbf{r})$ is the local solvent density at point **r**. It should be stressed that the correlation function that appears in eq. (A2) corresponds to the actual solvent density, which is determined not only by site *i* but by all neighboring sites as well (i.e., it is not the correlation function for an isolated group *i*). Thus, the contribution of this site to the solute–solvent energy depends on the nature of the surrounding groups and the conformation of the macromolecule.

In analogy to eq. (A1), the correlation function $g^{(1)}$ can be factored approximately into a product of group solvent correlation functions by the Kirkwood superposition approximation (KSA: Kirkwood, 1935); that is,

$$g^{(1)}(\mathbf{r}) = \prod_i g_{iw}(\mathbf{r}), \tag{A3}$$

where g_{iw} is the correlation function between the isolated site i and solvent. The solute–solvent entropy can then be separated into a sum of group contributions:

$$S_{sw} = -k \sum_i \rho \int g^{(1)}(\mathbf{r}) \ln g_{iw}(\mathbf{r}) \, d\mathbf{r}. \tag{A4}$$

When the KSA is not valid, one can write the general factorization as

$$g^{(1)}(\mathbf{r}) = \prod_i g_{iw}(\mathbf{r}) \prod_{ij} \delta g_{ijw}(\mathbf{r}) \prod_{ijk} \delta g_{ijkw}(\mathbf{r}) \cdots, \tag{A5}$$

and obtain for the entropy:

$$S_{sw} = -k \sum_i \rho \int g^{(1)}(\mathbf{r}) \ln g_{iw}(\mathbf{r}) \, d\mathbf{r} - k \sum_{ij} \rho \int g^{(1)}(\mathbf{r}) \ln \delta g_{ijw}(\mathbf{r}) \, \mathbf{r} - \cdots. \tag{A6}$$

In this case one can still split the solute–solvent entropy into group contributions by assigning half of the pair terms to each of the contributing atoms, and so on for the higher-order terms.

For the solvent reorganization energy and entropy one could use the expansion:

$$g^{(1)}(\mathbf{r}) - 1 = \sum_i [g_{iw} - 1] + \sum_{i \neq j} [g_{iw} - 1][g_{jw} - 1] + \sum_{i \neq j \neq k} [g_{iw} - 1][g_{jw} - 1][g_{kw} - 1] + \cdots, . \tag{A7}$$

where we abbreviate $g_{iw}(\mathbf{r})$ to g_{iw}. With this expansion we get a sum of terms that depend on one site, a sum of terms that depend on two sites, and so on. The terms get progressively smaller, because they correspond to increasing powers of a small quantity ($|a| > |a|^2 > |a|^3 > \ldots \to 0$ for $|a| < 1$).

Substituting from eq. (A7) into eqs. (36) and (38) gives

$$\Delta E_{ww} = \sum_i \tfrac{1}{2} \rho^2 \int g^{(1)}(\mathbf{r})[g_{iw} - 1] g^{(2)}(\mathbf{r}, \mathbf{r}') u_{ww}(\mathbf{r}, \mathbf{r}') \, d\mathbf{r} \, d\mathbf{r}' + \sum_{i \neq j} \tfrac{1}{2} \rho^2 \int g^{(1)}(\mathbf{r})[g_{iw} - 1][g_{jw} - 1] g^{(2)}(\mathbf{r}, \mathbf{r}') u_{ww}(\mathbf{r}, \mathbf{r}') \, d\mathbf{r} \, d\mathbf{r}' + \cdots, \tag{A8}$$

$$\Delta S_{ww} = -\sum_i \tfrac{1}{2} k\rho^2 \int g^{(1)}(\mathbf{r})[g_{iw} - 1]\{g^{(2)}(\mathbf{r}, \mathbf{r}') \ln g^{(2)}(\mathbf{r}, \mathbf{r}') - g^{(2)}(\mathbf{r}, \mathbf{r}') + 1\} \, d\mathbf{r} \, d\mathbf{r}' - \sum_{i \neq j} \tfrac{1}{2} k\rho^2 \int g^{(1)}(\mathbf{r})[g_{iw} - 1][g_{jw} - 1]\{g^{(2)}(\mathbf{r}, \mathbf{r}') \ln g^{(2)}(\mathbf{r}, \mathbf{r}') - g^{(2)}(\mathbf{r}, \mathbf{r}') + 1)\} \, d\mathbf{r} \, d\mathbf{r}' + \cdots \tag{A9}$$

By convention, we can split the two-site terms between the two sites (in the same way, that the interaction energy between two groups can be divided between them), the three-site terms equally among three sites, and so on. With eqs. (A2), (A6), (A8), and (A9) (and neglecting the PV term), we can express the solvation free energy of the solute as a sum of contributions from its groups:

$$\Delta G^{\text{slv}} = \sum_i \Delta G_i^{\text{slv}} \tag{A10}$$

with

$$\Delta G_i^{\text{slv}} = \rho \int g^{(1)}(\mathbf{r}) u_{iw}(\mathbf{r}) \, d\mathbf{r} + kT\rho \int g^{(1)}(\mathbf{r}) \ln g_{iw}(\mathbf{r}) \, d\mathbf{r} +$$

$$\tfrac{1}{2}\rho^2 \int g^{(1)}(\mathbf{r})[g_{iw} - 1]g^{(2)}(\mathbf{r}, \mathbf{r}') u_{ww}(\mathbf{r}, \mathbf{r}') \, d\mathbf{r} \, d\mathbf{r}' +$$

$$\tfrac{1}{2}\sum_j \tfrac{1}{2}\rho^2 \int g^{(1)}(\mathbf{r})[g_{iw} - 1][g_{jw} - 1]g^{(2)}(\mathbf{r}, \mathbf{r}') u_{ww}(\mathbf{r}, \mathbf{r}') \, d\mathbf{r}' \, d\mathbf{r}' + \cdots +$$

$$\tfrac{1}{2}kT\rho^2 \int g^{(1)}(\mathbf{r})[g_{iw} - 1]\{g^{(2)}(\mathbf{r}, \mathbf{r}') \ln g^{(2)}(\mathbf{r}, \mathbf{r}') - g^{(2)}(\mathbf{r}, \mathbf{r}') + 1\} \, d\mathbf{r} \, d\mathbf{r}' +$$

$$\tfrac{1}{2}\sum_j \tfrac{1}{2}kT\rho^2 \int g^{(1)}(\mathbf{r})[g_{iw} - 1][g_{jw} - 1]\{g^{(2)}(\mathbf{r}, \mathbf{r}') \ln g^{(2)}(\mathbf{r}, \mathbf{r}') - g^{(2)}(\mathbf{r}, \mathbf{r}') + 1\}$$

$$d\mathbf{r} \, d\mathbf{r}' + \cdots. \tag{A11}$$

Acknowledgments Partial support of this work was provided by an RCMI grant from the National Institutes of Health to CCNY (T.L.) and by a grant from the National Institutes of Health to Harvard University (M.K.).

References

Alexov, E. G., and Gunner, M. R. (1997) *Biophys. J.* **74**, 2075–93.
Anfinsen, C. B. (1973) Science **181**, 223–30.
Archontis, G., and Karplus, M. (1996) *J. Chem. Phys.* **105**, 11246–60.
Arnold, F. H. (1998) *Acc. Chem. Res.* **31**, 125–31.
Baker, D. (1998) *Nature Struct. Biol.* **5**, 1021–4.
Baker, D., and Agard, D. A. (1994) *Biochemistry* **33**, 7505–9.
Baldwin, R. L. (1986) *Proc. Natl. Acad. Sci. USA* **83**, 8069–72.
Bashford, D., and Karplus, M. (1990) *Biochemistry* **29**, 10219–25.
Ben-Naim, A. (1978) *J. Phys Chem.* 82, 792–803.
Beveridge, D. L., and DiCapua, F. M. (1989) *Ann. Ref. Biophys. Chem.* **18**, 431–92.
Boczko, E.M., and Brooks, C. L. I. (1995) *Science* **269**, 393–6.
Bond, C. J., Wong, K.-B., Clarke, J., Fersht, A. R., and Daggett, V. (1997) *Proc. Natl. Acad. Sci. USA* **94**, 13409–13.
Boresch, S., Archontis, G., and Karplus, M. (1994) *Proteins* **20**, 25–33.
Boresch, S., and Karplus, M. (1995) *J. Mol. Biol.* **254**, 801–7.
Brady, G. P., and Sharp, K. A. (1995) *J. Mol. Biol.* **254**, 77–85.
Brady, G. P., Szabo, A., and Sharp, K. A. (1996) *J. Mol. Biol.* **263**, 123–5.

Brandts, J. F. (1964) *J. Am. Chem. Soc.* **86**, 4302–14.
Brooks, B., and Karplus, M. (1983) *Proc. Natl. Acad. Sci. USA* **80**, 6571–5.
Brooks, B. R., Bruccoleri, R. E., Olafson, B. D., States, D. J., Swaminathan, S., and Karplus, M. (1983) *J. Comput. Chem.* **4**, 187–217.
Brooks, C. L., III, Karplus, M., and Pettitt, B. M. (1988) Proteins: a theoretical perspective of dynamics, structure, and thermodynamics. *Adv. Chem. Phys.* **71**.
Brooks, C. L. I. (1990) Molecular simulations of protein structure, dynamics and thermodynamics. In Computer Modeling of Fluids Polymers and Solids, C.R.A. Catlow, S. Parker, and M. Allen (eds.). Dordrecht: Kluwer Academic Publishers.
Caflisch, A., and Karplus, M. (1995) *J. Mol. Biol.* **252**, 672–708.
Colonna-Cesari, F., and Sander, C. (1990) *Biophys. J.* **57**, 1103–7.
de Groot, S. R., and Mazur, P. (1984) *Non-equilibrium thermodynamics.* New York: Dover.
Devoe, H. (1969) Theory of the conformations of biological macromolecules in solution. In: *Structure and Stability of Biological Macromolecules*, S. N. Timascheff and G. D. Fasman (eds.). New York: Dekker.
Dill, K. A. (1997) *J. Biol. Chem.* **272**, 701–4.
Dill, K. A., and Shortle, D. (1991) *Ann. Rev. Biochem.* **60**, 795.
Dinner, A. R., and Karplus, M. (1998) *Nature Str. Biol.* **5**, 236–40.
Eigen, M. (1986) *Chem Scr. B 26*, 13–26.
Eisenberg, D., and McLachlan, A. D. (1986) *Nature* **319**, 199–203.
Eisenberg, D. S., and Crothers, D. M. (1979) *Physical Chemistry: with Applications to the Life Sciences.* Menlo Park, Calif.: Benjamin/Cummings Pub. Co.
Elber, R., and Karplus, M. (1987) *Science* **235**, 318–21.
Ellis and Hartl (1999) *Curr. Opin. Struct. Biol.* **9**, 102–10.
Eriksson, A. E., Baase, W. A., Zhang, Z.J., Heinz, D. W., Blaber, M., Baldwin, E. P., and Matthews, B. W. (1992) *Science* **255**, 178–183.
Ernst, J. A., Clubb, R. T., Zhou, H.-Z., Gronenborn, A. M., and Clore, G. M. (1995) *Science* **267**, 1813–16.
Fersht, A. R. (1987) *TIBS* **12**, 301–4.
Fersht, A. R. (1988) *Biochem* **27**, 1577–1580.
Fersht, A. R. (1999) *The Mechanism of Enzyme Catalyses and Protein Folding.* W. H. Freeman.
Field, M. J., Bash, P. A., and Karplus, M. (1990) *J. Comp. Chem.* **11**, 700–33.
Fisher, H. F., and Singh, N. (1995) *Methods Enzymol.* **259**, 194–221.
Fraternali, F., and van Gunsteren, W. F. (1996) *J. Mol. Biol.* **256**, 939–48.
Frauenfelder, H., Sligar, S. G., and Wolynes, P. G. (1991) *Science* **254**, 1598–1603.
Gao, J., Kuczera, K., Tidor, B., and Karplus, M. (1989) *Science* **244**, 1069–72.

Gerber, P. R., Mark, A. E., and van Gunsteren, W. F. (1993) *J. CAMD* **7**, 305.
Glansdorff, P., and Prigogine, I. (1971) *Thermodynamic Theory of Structure, Stability and Fluctuations*. London: Wiley-Interscience.
Goldberg, M. E. (1985) *TIBS* **10**, 388–91.
Gomez, J., Hilser, V. J., Xie, D., and Freire, E. (1995) *Proteins* **22**, 404.
Hagler, A. T., Huler, E., and Lifson, S. (1974) *J. Am. Chem. Soc.* **96**, 5319–27.
Halle, B., Andersson, T., Forsen, S., and Lindman, B. (1981) *J. Am. Chem. Soc.* **103**, 500–8.
Hansen, J. P., and McDonald, I. R. (1986) *Theory of Simple Liquids*. London: Academic Press.
Herschbach, D. R., Johnston, H. S., and Rapp, D. (1959) *J. Chem. Phys.* **31**, 1652–61.
Hirata, E., and Rossky, P. J. (1981) *Chem. Phys. Lett.* **83**, 329–34.
Honig, B., and Nicholls, A. (1995) *Science* **268**, 1144–9.
Honig, B., and Yang, A.-S. (1995) *Adv. Prot. Chem.* **46**, 27–58.
Ichiye, T., and Chandler, D. (1988) *J. Phys. Chem.* **92**, 5257–61.
Jackson, S. E., and Fersht, A. R. (1991) *Biochemistry* **30**, 10428–35.
Jain, M. K. (1988) *Introduction to Biological Membranes*. New York: Wiley-Interscience.
Jorgensen, W. L., Chandrasekhar, J., Madura, J. D., Impey, R. W., and Klein, M. L. (1983) *J. Chem. Phys.* **79**, 926–35.
Jou, D., and Llebot, J. E. (1990) *Introduction to the Thermodynamics of Biological Processes*. Englewood Cliffs: Prentice Hall.
Kang, Y. K., Nemethy, G., and Scheraga, H. A. (1987) *J. Chem. Phys.* **91**, 4105–9.
Karplus, M., Ichiye, T., and Pettitt, B. M. (1987) *Biophys. J.* **52**, 1083–5.
Katchalsky, A., and Curran, P. D. (1965) *Nonequilibrium Thermodynamics in Biophysics*. Cambridge, Harvard University Press.
Kauffman, S. A. (1993) *The Origins of Order*. Oxford University Press.
Kauzmann, W. (1959) *Adv. Protein Chem.* **1**, 14–63.
Kazmirsky, S. L., and Daggett, V. (19987) *J. Mol. Biol.* **277**, 487–506.
Kellis, J. T., Jr., Nyberg, K., and Fersht, A. R. (1989) *Biochemistry* **28**, 4914–22.
Kirkwood, J. G. (1935) *J. Chem. Phys.* **3**, 300–13.
Kollman, P. A. (1993) *Chem. Rev.* **93**, 2395–2418.
Lattman, E. E., Fiebig, K. M., and Dill, K. A. (1994) *Biochemistry* **33**, 6158–66.
Lau, F. T. K., and Karplus, M. (1994) *J. Mol. Biol.* **236**, 1049–66.
Lazaridis, T. (1998a) *J. Phys. Chem.* **102**, 3531–41.
Lazaridis, T. (1998b) *J. Phys. Chem.* **102**, 3542–50.
Lazaridis, T., Archontis, G., and Karplus, M. (1995) *Adv. Prot. Chem.* **47**, 231–306.
Lazaridis, T., and Karplus, M. (1996) *J. Chem. Phys.* **105**, 4294–4316.

Lazaridis, T., and Karplus, M. (1997) *Science* **278**, 1928–31.
Lazaridis, T., and Karplus, M. (1999a) *J. Mol. Biol.* **288**, 477–87.
Lazaridis, T., and Karplus, M. (1999b) *Proteins* **35**, 133–52.
Lazaridis, T., and Karplus, M. (1999c) *Biophys. Chem.* **78**, 2097–17
Lazaridis, T. (2000) *J. Phys. Chem. B* **104**, 4964–79.
Lazaridis, T., Lee, I., and Karplus, M. (1997) *Protein Sci*, **6**, 2589–605.
Lazaridis, T., and Paulaitis, M. E. (1992) *J. Phys. Chem.* **96**, 3847–55.
Lazaridis, T., and Paulaitis, M. E. (1994) *J. Phys. Chem.* **98**, 635–42.
Lee, B. (1993) *Prot. Sci.* **2**, 733–8.
Lum, K., Chandler, D., and Weeks, J. D. (1999) *J. Phys. Chem. B* **103**, 4570–7.
MacKerell, A. D., Jr., Bashford, D., ..., and Karpalus, M. (1998) *J. Phys. Chem. B* **102**, 3586–616.
Makhatadze, G. I., and Privalov, P. L. (1990) *J. Mol. Biol.* **213**, 375–84.
Makhatadze, G. I., and Privalov, P. L. (1993) *J. Mol. Biol.* **232**, 639–69.
Makhatadze, G. I., and Privalov, P. L. (1994) *Biophys. Chem.* **51**, 291–309.
Makhatadze, G. I., and Privalov, P. L. (1995) *Adv. Prot. Chem.* **47**, 307–425.
Mark, A. E. and van Gunsteren, W. F. (1994) *J. Mol. Biol.* **240**, 167–76.
Matsumura, M., Becktel, W. J., and Matthews, B. W. (1988) *Nature* **334**, 406–10.
Matubayasi, N., Reed, L. H., and Levy, R. M. (1994) *J. Phys. Chem.* **98**, 10640–9.
McCammon, J. A., and Harvey, S. C. (1987) *Dynamics of Proteins and Nucleic Acids*. Cambridge: Cambridge University Press.
McQuarrie, D. A. (1976). *Statistical Mechanics*. New York: Harper & Row.
Miller, S., Janin, J., Lesk, A. M., and Chothia, C. (1987). *J. Mol. Biol.* **196**, 641–56.
Murphy, K. P., and Freire, E. (1992) *Adv. Protein Chem.* **43**, 313–61.
Murphy, K. P., and Gill, S. J. (1991) *J. Mol. Biol.* **222**, 699–709.
Myers, J. K., and Pace, C. N. (1996) *Biophys. J.* **71**, 2033–9.
Neria, E., Fischer, S., and Karplus, M. (1996) *J. Chem. Phys.* **105**, 1902–21.
Nölting, B., Golbik, R., Soler-González, A. S., and Fersht, A. R. (1997) *Biochemistry* **36**, 9899–9905.
Ooi, T., Oobatake, M., Nemethy, G., and Scheraga, H. A. (1987) *Proc. Natl. Acada. Sci.* **84**, 3086–90.
Otting, G., Liepinsh, E., and Wüthrich, K. (1991) *Science* **254**, 974–80.
Petitt, B. M., and Rossky, P. J. (1986) *J. Chem. Phys.* **84**, 5836–44.
Prevost, M., Wodak, S. J., Tidor, B., and Karplus, M. (1991) *Proc. Natl. Acad. Sci. USA* **88**, 10880–4.
Prigogine, I. (1961) *Introduction to the Thermodynamics of Irreversible Processes*. New York: Interscience.
Privalov, P. L. (1979) *Adv. Prot. Chem.* **33**, 167–241.
Privalov, P. L., and Gill, S. J. (1988) *Adv. Protein Chem.* **39**, 191–234.
Privalov, P. L., and Makhatadze, G. I. (1992) *J. Mol. Biol.* **224**, 715–23.

Privalov, P. L., Tiktopoulo, E. I., Venyaminov, S. Y., Griko, Y. V., Makhatadze, G. I., and Khechinashvili, N. N. (1989) *J. Mol. Biol.* **205**, 737–50.
Ptitsyn, O. B. (1995) *Adv. Prot. Sci.* **47**, 83–229.
Rashin, A. A. (1993) *Prog. Biophys. Mol. Biol.* **60**, 73–200.
Reiher, W. E., III (1985) *Theoretical Studies of Hydrogen Bonding.* Ph.D. Thesis, Harvard University.
Reiss, H. (1965) *Adv. Chem. Phys.* **9**, 1–84.
Sandberg, W. S., and Terwilliger, T. C. (1991) *Proc. Natl. Acad. Sci. USA* **88**, 1706–10.
Schaefer, M., Bartels, C., and Karplus, M. (1998) *J. Mol. Biol.* **284**, 835–48.
Schaefer, M., and Karplus, M. (1996) *J. Phys. Chem.* **100**, 1578–9.
Schrödinger, E. (1944) *What is life?* Cambridge, Cambridge University Press.
Serrano, L., Kellis, J. T., Jr., Cann, P., Matouschek, A., and Fersht, A. R. (1992) *J. Mol. Biol.* **224**, 783–804.
Sharp, K., and Honig, B. (1990) *Ann. Rev. Biophys. Biophys. Chem.* **19**, 301–32.
Shirley, B. A., Stanssens, P., Hahn, U., and Pace, C. N. (1992) *Biochemistry* **31**, 725–32.
Shortle, D. (1996) *FASEB J.* **10**, 27–34.
Shortle, D., Chan, H. S., and Dill, K. A. (1992) *Prot. Science* **1**, 201–15.
Shortle, D., Stites, W. E., and Meeker, A. K. (1990) *Biochemistry* **29**, 8033–41.
Simonson, T., and Brünger, A. T. (1992) *Biochemistry* **31**, 8661–74.
Sitkoff, D., Sharp, K. A., and Honig, B. (1994) *J. Phys. Chem.* **98**, 1978–88.
Sneddon, S. F., and Tobias, D. J. (1992) *Biochemistry* **31**, 2842–6.
Spolar, R. S., Ha, J.-H., and Record, M. T., Jr. (1989) *Proc. Natl. Acad. Sci. USA* **86**, 8382–5.
Spolar, R. S., Livingstone, J. R., and Record, M. T., Jr. (1992) *Biochemistry* **31**, 3947–55.
Spolar, R. S., and Record, M. T., Jr. (1994) *Science* **263**, 777–84.
Still, W. C., Tempczyk, A., Hawley, R. C., and Hendrickson, T. (1990) *J. Am. Chem. Soc.* **112**, 6127–9.
Stouten, P. F. W., Frommel, C., Nakamura, H., and Sander, C. (1993) *Mol. Simul.* **10**, 97–120.
Straatsma, T. P., and McCammon, J. A. (1992) *Ann. Rev. Phys. Chem.* **95**, 1175–88.
Sturtevant, J. M. (1977) *Proc. Natl. Acad. Sci. USA* **74**, 2236–40.
Tembe, B., and McCammon, J. A. (1984) *Comput. & Chem* **8**, 281–283.
van Vlijmen, H. W. T., Schaefer, M. and Karplus, M. (1998) *Proteins* **33**, 145–58.
Velicelebi, G., and Sturtevant, J. M. (1979) *Biochemistry* **18**, 1180–6.
Weiner, S. J., Kollman, P. A., Nguyen, D. T., and Case, D. A. (1986) *J. Comput. Chem.* **7**, 230–52.

Wells, T. N. C., and Fersht, A. R. (1986) *Biochemistry* **25**, 1881–6.
Wesson, L., and Eisenberg, D. (1992) *Protein Science* **1**, 227–35.
Wiley, D. C., and Skehel, J. J. (1987) *Ann. Rev. Biochem.* **56**, 365–94.
Yang, A.-S., Gunner, M. R., Sampogna, R., Sharp, K., and Honig B. (1993) *Proteins* **15**, 252–65.
Yang, A.-S., and Honig, B. (1992) *Curr. Opin. Struct. Biol.* **2**, 400-5.
Yu, H.-A., and Karplus, M. (1988) *J. Chem. Phys.* **89**, 2366–79.
Yu, H.-A., Pettitt, B. M., and Karplus, M. (1991) *J. Am. Chem. Soc.* **113**, 2425–34.
Zotin, A. I. (1972) Thermodynamic Aspects of Developmental Biology. Basel: S. Karger.

2

Thermodynamic Dissection of Cooperativity in Ligand Recognition

Thierry Rose and Enrico Di Cera

A large body of experimental data suggests that biological macromolecules accomplish diverse functions using cooperative interactions among structural domains. The best-known example of cooperativity is offered by hemoglobin, in which binding of oxygen to one heme affects the binding properties of other hemes in the molecule, and oxygen release to the tissues is allosterically controlled by the uptake of protons and organic phosphates at other sites (Perutz, 1989). Cooperativity, however, is not limited to ligand-binding processes and allosteric proteins. It is an inherent component of protein stability, providing the necessary communication among residues of the protein to maintain the folded structure (Creighton, 1992). It is also a key player in determining secondary structure, as illustrated by the helix–coil transitions of biopolymers (Zimm & Bragg, 1959). More recently, systematic mutagenesis studies of binding epitopes have fostered the notion that cooperativity may be a fundamental ingredient of any recognition event (Di Cera, 1998a,b).

Thermodynamics provides rigorous and model-independent tools to decipher cooperativity at the molecular level. The thermodynamic approach is both simple in its formulation and general in its applicability and implications. In this chapter, we examine three basic aspects of cooperativity in ligand recognition—a key phenomenon taking place in practically every biological system—and focus on the contribution of individual amino-acid residues to ligand binding. First, we demonstrate that amino-acid residues forming binding epitopes in a protein tend to participate in ligand binding via nonadditive effects, as revealed by Ala-scanning mutagenesis studies. Then, we show how substrate binding by an enzyme can be decomposed in terms of the nonadditive contributions of recognition subsites. Finally, we use structural information to predict the nonadditive contributions to substrate binding by an enzyme, and compare the results with available experimental data. These examples are meant to illustrate the power of a thermodynamic approach to ligand recognition that integrates structural and functional information and is of general applicability to studies of biological specificity.

1. Non-additive Contributions to Ligand Binding

1.1 Ala-scanning Mutagenesis and the Definition of Epitopes

Central to many areas of investigation in structural biology and biochemistry is the identification of epitopes for ligand recognition. The structural analysis focuses on the identification of residues that contact the ligand in the complex. This analysis complements a thermodynamic approach aimed at identifying residues that are energetically linked to the binding event. Identification of these residues then raises the question of whether they act independently or cooperatively and, if the latter, what is the origin of cooperativity.

Binding epitopes are usually mapped energetically by replacing residues that structural information or sequence analysis indicates as being involved in ligand recognition. There are twenty possible choices for any given residue in a protein; therefore, if one specific residue is to be replaced, there are nineteen possibilities. In practice, the residue of choice for the replacement is Ala. The rationale behind Ala-scanning mutagenesis is that all interactions of a side chain, except for the C_β atom, are eliminated (Lau and Fersht, 1987; Cunningham and Wells, 1989). The effect of the mutation is assessed from the difference in binding free energy between the mutant and wild type as

$$\Delta \Delta G_{mut} = \Delta G_{mut} - \Delta G_{wt}. \qquad (1)$$

The expectation in this approach is that, when the Ala substitution is made at sites important for ligand recognition, it will destabilize the form of the macromolecule in complex with the ligand without changing the properties of the free form. The resulting loss in binding affinity, or $\Delta \Delta G_{mut} > 0$, can then be used to assess the energetic contribution of the mutated residue to ligand binding. Values of $\Delta \Delta G_{mut}$ can be used to color-code the surface of the macromolecule and to visually display the spatial organization of the binding epitope and areas energetically sensitive to ligand recognition (Clackson and Wells, 1995). This strategy has proved invaluable in the study of a number of systems of biological importance. Ala-scanning mutagenesis has found myriad applications in the identification and energetic characterization of structural epitopes recognizing specific ligands (Tsiang et al., 1995; Dickinson et al., 1996; Pakianathan et al., 1997), or the structural determinants of protein stability (Green et al., 1992; Horovitz and Fersht, 1992; Fersht and Serrano, 1993; Matthews, 1993; Yu et al., 1995; Shortle, 1996) and enzyme mechanism (Carter and Wells, 1988). However, there are potential pitfalls in this strategy that must be recognized in practical applications (Greenspan and Di Cera, 1999).

It is reasonable to expect that mutation of surface residues will significantly perturb formation of the complex without compromising the properties of the free form of the macromolecule. However, the binding free energy used as a measure of the energetic perturbation of the Ala replacement (eq. (1)) cannot sort out effects on the complex and free species. A positive value of $\Delta\Delta G_{mut}$ means that the mutation has reduced the stability of the complex more than that of the free form, and does not imply necessarily that the residue mutated to Ala stabilizes the complex. An Ala substitution that stabilizes the free form of the macromolecule, but has no effect on the bound form, will decrease the binding affinity and result in $\Delta\Delta G_{mut} > 0$. In this case, however, the residue replaced by Ala will be assigned to the binding epitope for the ligand and deemed important for recognition, while in fact it is only important for the stability of the free form of the macromolecule, independent of any ligand. For example, Ala substitutions of R221a, K224, and Y225 in thrombin affect fibrinogen recognition up to 100-fold (Dang et al., 1997a; Guinto et al., 1999), but none of these residues makes contract with fibrinogen in the crystal structure (Stubbs et al., 1992) (fig. 2.1). In the absence of structural information, these residues would be assigned mistakenly to the binding epitope (as defined by intermolecular contact) for fibrinogen. On the other hand, Ala replacement of R173 has only a modest effect on fibrinogen binding (Guinto et al., 1995; Tsiang et al., 1995). In the absence of structural information, R173 would not be included in the epitope for fibrinogen binding, although it actually makes an ion-pair interaction with E11 of the fibrinogen Aα chain (Stubbs et al., 1992).

A more fundamental limitation of single-site Ala replacements is that they neglect *a priori* the contribution of possible site–site interactions to ligand recognition. Results from studies where the importance of site–site interactions in mutational effects has been addressed experimentally have fostered the unfortunate notion that residues tend to participate independently in stability and recognition (Sandberg and Terwilliger, 1989; Shirley et al., 1989; Wells, 1990) and that interactions only occur among residues close in space (Carter et al., 1984; Carter and Wells, 1988; Horovitz and Fersht, 1992; Mildvan et al., 1992; Wells, 1990). It has now been recognized that interactions may involve residues as far as 3 nm away from each other (Shortle and Meeker, 1986; Perry et al., 1989; Howell et al., 1990; LiCata et al., 1990; Scrutton et al., 1990; Green and Shortle, 1993; Jackson and Fersht, 1993; Robinson and Sligar, 1993; LiCata and Ackers, 1995; Dill, 1997). Hence, there is good reason to believe that interactions are present in nearly every system and provide the most important ingredient to ligand recognition.

52 THERMODYNAMICS IN BIOLOGY

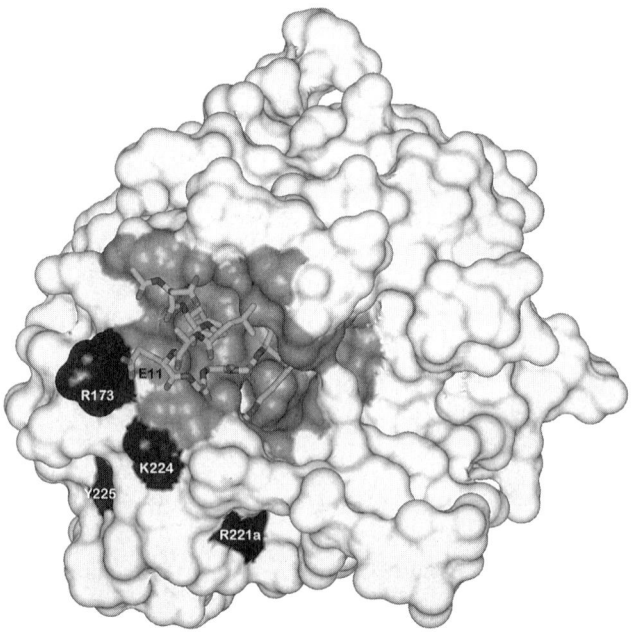

Figure 2.1. Surface of human thrombin (white) bound to a fragment of the fibrinogen Aα chain (stick model), according to the crystal structure (Stubbs et al., 1992). The surface in grey depicts the binding epitope for fibrinogen defined as the set of thrombin residues making contacts with the ligand (i.e., within 0.5 nm of the ligand). Residues in black have been subject to Ala substitution. Mutation of R173 has only a modest effect on fibrinogen binding (Guinto et al., 1995), although this residue makes an ion-pair interaction with E11 of the fibrinogen Aα chain and is part of the binding epitope. On the other hand, mutation of R221a, K224, and Y225 affect significantly, up to 100-fold, fibrinogen binding (Dang et al., 1997a: Guinto et al., 1999). None of these residues is part of the epitope defined by intermolecular contact as determined from the crystal structure.

A compelling argument in favor of the existence of cooperativity in ligand recognition is as follows. The key assumption of Ala-scanning mutagenesis is that the Ala replacement has the only effect of eliminating the interactions of the side chain beyond the C_β (Lau and Fersht, 1987; Cunningham and Wells, 1989). If this assumption is tenable, and if the Ala replacement is an unbiased probe of the energetic contribution of a given residue to binding, then the Ala mutation at any position of the epitope should convert the free-energy contribution to zero. If this is not the case, then the Ala replacement has introduced new properties at the site, thereby invalidating the assumption. Hence, a binding epitope composed of independent residues should be such that the sum of the free-energy changes due to Ala replacement over all sites in the epitope, with changed sign, approximates the actual free energy of binding measured experimentally for the wild type. Inspection of the results in table 2.1 for a number of systems studied to date by Ala-scanning mutagenesis shows that this is not the case.

A large discrepancy exists between the calculated and experimentally determined values. In the case of human growth hormone and insulin binding to their receptors, or the peptide QL9 binding to the T-cell receptor (Manning et al., 1998), the binding affinity calculated from the results of the Ala scan is greatly overestimated. In the case of BPTI binding to trypsin (Castro and Anderson, 1996), tissue factor binding to coagulation factor VIIa (Dickinson et al., 1996), or linolenate binding to intestinal fatty-acid-binding protein (Richieri et al., 1997), the binding affinity is grossly underestimated. When the affinity is underestimated, it may be argued that important interactions might have been missed in the Ala scan. This is unlikely in the case of the interaction of tissue factor with VIIa, where 112 residues were targeted by mutagenesis, or for intestinal fatty-acid-binding protein, where 23 important residues in the binding cavity were replaced. On the other hand, when the affinity is overestimated, it may be argued that the Ala scan might have included sites of marginal importance. Again, this is unlikely the case in the interaction of human growth hormone with its receptor, or for the peptide QL9 with the T-cell receptor, where the functional epitope is a small hot spot. The results of intestinal fatty-acid-binding protein are particularly instructive in so far as they show that the discrepancy between calculated and experimentally determined values depends on the particular ligand examined. The difference changes from −11.5 kcal/mol for linolenate to 1.4 kcal/mol for stearate. Given the comparable size of the fatty acids listed in table 2.1 and their comparable binding affinity, this large difference cannot be due to intrinsic properties of the ligand. Rather, it suggests the presence of communication among the protein residues that is sensitive to the particular ligand bound. In the case of insulin binding to its receptor, the affinity is underestimated when 26 residues are mutated to Ala (Kristensen et al., 1997). However, when the results are combined with other Ala scans under identical conditions (Cosmatos et al., 1978; Marki et al., 1979; Kobayashi et al., 1984; Mirmira and Tager, 1991; Nakagawa and Tager, 1991, 1992) to cover a total of 38 residues, the affinity is grossly overestimated. A similar situation is encountered in the binding of a monoclonal antibody to coagulation factor VIII (Lubin et al., 1997). Again, when the Ala scan involves most of the residues responsible for binding, a large discrepancy is seen between calculated and experimentally determined values for the binding of the ligand to the wildtype, underscoring the important role played by interactions among residues in the recognition process.

Table 2.1. Comparison of Free-energy Values (in kcal/mol) for Ligand Recognition Measured Experimentally and Calculated from Single-site Ala Scans

System	Ala replacements	ΔG_{cal}	ΔG_{exp}	$\delta\Delta G_{coop}$	Reference(s)
hGH-hGHbp[a]	30	−25.9	−12.3	13.6	Clackson & Wells, 1995
BPTI-chymotrypsin	15	−6.4	−10.7	−4.3	Castro & Anderson, 1996
VIIa-TF[b]	112	−9.7	−15.4	−5.7	Dickinson et al., 1996
I-FABP[c] (palmitate)	23	−6.8	−10.9	−4.1	Richieri et al., 1997
I-FABP[c] (stearate)	23	−13.1	−11.7	1.4	Richieri et al., 1997
I-FABP[c] (oleate)	23	−8.5	−10.7	−2.2	Richieri et al., 1997
I-FABP[c] (linoleate)	23	−5.4	−10.0	−4.6	Richieri et al., 1997
I-FABP[c] (linolenate)	23	2.4	−9.1	−11.5	Richieri et al., 1997
I-FABP[c] (arachidonate)	23	−3.6	−9.5	−5.9	Richieri et al., 1997
GCSF-GCSF receptor[d]	27	−14.5	−11.3	3.2	Young et al., 1997
RANTES-CCR1[e]	16	−5.2	−12.3	−7.1	Pakianathan et al., 1997
RANTES-CCR3[e]	16	−10.5	−13.0	−2.5	Pakianathan et al., 1997
Insulin-receptor	26	−8.9	−11.2	−2.3	Kristensen et al., 1997
Insulin-receptor	38	−23.4	−11.2	12.1	Cosmatos et al., 1978; Marki et al., 1979; Kobayashi et al., 1984; Mirmira & Tager, 1991; Nakagawa & Tager, 1991, 1992; Kristensen et al., 1997
VII-mAb413[f]	10	−16.6	−13.6	3.0	Lubin et al., 1997
TrB₂-mAb164[g]	9	−17.4	−13.5	3.9	Rondard & Bedouelle, 1998
TCR-QL9[h]	40	−16.6	−9.5	7.1	Manning et al., 1998

ΔG_{calc}, ΔG_{exp}, and $\delta\Delta G_{coop}$ are defined in eq. (2).

[a] Human growth hormone (hGH) binding to the extracellular domain of its first bound receptor (hGHbp).
[b] Tissue factor (TF) binding to coagulation factor VIIa.
[c] Intestinal fatty-acid-binding protein.
[d] Granulocyte-colony stimulating factor (GCSF).
[e] CC-chemokine regulated upon activation normal T-cell expressed and secreted interacting with its receptors.
[f] Monoclonal antibody 413 binding to coagulation factor VIII.
[g] Monoclonal antibody 164 binding to tryptophan repressor B₂.
[h] Nonamer peptide QL9 binding to T-cell receptor (TCR)

An epitope containing all residues replaced by Ala should bind a ligand with a null ΔG if the residues act independently. A binding free energy of zero means that the ligand experiences no net energetic change in the standard state in going from the free to the bound state, and that the all-Ala binding epitope is energetically neutral. Although this scenario may seen paradoxical, its validity within reasonable energetic terms is key to the approach based on Ala scans. If the discrepancies in table 2.1 are the result of specific favorable or unfavorable contributions to recognition introduced by the presence of Ala at any given site, the assignment of epitopes with Ala-scanning mutagenesis becomes context-dependent and invalid. It is possible that Ala replacements introduce additional properties at the site of mutation, and that these properties bias the energetic balance of the substitution. However, this bias is likely to be small. We propose that the large discrepancies in table 2.1 are due to the neglect of energetic contributions arising from site–site interactions that cannot be quantified by single-site Ala scans.

An approximate measure of the extent of interactions among residues is given by the difference, $\delta\Delta G_{coop}$, between the experimentally determined, ΔG_{exp}, and calculated, ΔG_{calc}, values of the free energy of binding:

$$\delta\Delta G_{coop} = \Delta G_{exp} - \Delta G_{calc} = \Delta G_{wt} + \sum_{j=1}^{N}({}^{j}\Delta G_{mut} - \Delta G_{wt})$$

$$= \Delta G_{wt} + \sum_{j=1}^{N}{}^{j}\Delta\Delta G_{mut}. \qquad (2)$$

The value of ΔG_{exp} is the same as the free energy of the wild type, ΔG_{wt}. The calculated ΔG_{calc} is the sum of the differences between the free energy of the mutant and wild type for all N mutants in the epitope, with changed sign. In the absence of interactions among the residues being mutated to Ala, and under the assumption that the Ala substitution is energetically neutral, $\delta\Delta G_{coop}$ should be as close as possible to zero. Hence,

$$\Delta G_{wt} = -\sum_{j=1}^{N}{}^{j}\Delta\Delta G_{mut} \qquad (3)$$

is the expected result for an epitope composed of independent residues.

It should be pointed out that ΔG_{wt} depends on the definition of the standard state for the thermodynamic equilibrium (Janin, 1995)

$$\Delta G_{wt} = RT \ln K_d - RT \ln c_0, \qquad (4)$$

where R is the gas constant, T the absolute temperature, K_d the equilibrium dissociation constant and, c_0 the concentration (in molar units) used to define the standard state. The convention is to assume $c_0 = 1$

M, but the arbitrariness of c_0 makes the absolute value of ΔG_{wt} not uniquely defined. On the other hand, the values $\Delta\Delta G_{mut}$ in eq. (3) do not depend on c_0 and therefore are absolutely defined. This implies

$$RT\ln K_d + \sum_{j=1}^{N}{}^j\Delta\Delta G_{mut} = \delta\Delta G_{coop} + RT\ln c_0. \quad (5)$$

If the values of K_d for the wild type are used to calculate ΔG_{wt}, then (in principle) the values of $\delta\Delta G_{coop}$ are only defined minus an arbitrary term $RT\ln c_0$, and there is no a priori reason why they should be zero even in the absence of interactions. However, because c_0 is constant, the values of $\delta\Delta G_{coop}$ should be the same for any system studied, in the absence of interactions. Again, inspection of table 2.1 shows that this is not the case, and that the nonzero values of $\delta\Delta G_{coop}$ are not the result of the arbitrary standard state defining the binding free energy ΔG_{wt}.

The presence of interactions invalidates the energetic assignments derived from single-site Ala scans, because the contribution of a given residue to ligand binding will depend on the state (wild type or mutated) of other residues. The extent to which interactions affect the assignments based on single-site Ala scans must be evaluated in each case, and this complicates the identification of epitopes (Greenspan & Di Cera, 1999). The contribution of a residue in the presence of cooperativity involves effects of multiple order. A first-order contribution comes from contacts made directly with the ligand. Higher-order contributions may come from the coupling between the residue and other structural components. The residue recognizing the ligand may be involved in a number of interactions with other residues via short-range van der Waals coupling, long-range electrostatic coupling, or large-scale conformational transitions. For interactions of second order, the construction of double mutations becomes necessary to assess the energetic contribution to stability and ligand recognition, and so forth for higher-order interactions. If an epitope contains N residues, a complete single-site Ala scan requires N mutations, and a double-site Ala scan requires $N(N-1)/2$ mutations. The problem of correctly assessing the energetic contribution of residues in a functional epitope using site-directed mutagenesis is combinatorially challenging, and demands elucidation of the site–site coupling patterns.

1.2. Double-mutant Cycles

Consider the general case of a system composed of N sites that can exist in two states: 0 (wild type) and 1 (mutant). For this system ΔG_j is the free energy change associated with the $0 \to 1$ transition at site j when all other sites are in state 0. This term is the difference in free energy between the configuration with site j perturbed and the wild type, resulting in the loss

($\Delta G_j > 0$) or gain ($\Delta G_j < 0$) of specificity, at equilibrium or in the transition state, due to perturbation of that site. There are N such terms to be taken into account, one for each site. Consider, then, the double perturbation at sites i and j. The free energy change for such perturbation can be written as the sum $\Delta G_i + \Delta G_j + \Delta G_{ij}$, where ΔG_{ij} is the interaction free energy between sites i and j when the perturbation is applied at both sites. This ΔG_{ij} is the same as the coupling free energy in the thermodynamic cycle (Ackers and Smith, 1985; Horovitz and Fersht, 1990; Wells, 1990; Di Cera, 1998a,b)

$$\begin{array}{ccc} & \Delta G_i & \\ M_{00} & \Leftrightarrow & M_{10} \\ \Delta G_j \updownarrow & & \updownarrow \; \Delta G_j + \Delta G_{ij}, \\ M_{01} & \Leftrightarrow & M_{11} \\ & \Delta G_i + \Delta G_{ij} & \end{array} \qquad (6)$$

where the suffix denotes the state—wild type or mutant—of sites i and j of the macromolecule M. A negative value of ΔG_{ij} indicates positive coupling between the perturbations at sites i and j in enhancing specificity, or negative coupling in reducing it, and vice versa for a positive value. A null value of ΔG_{ij} indicates the absence of coupling between the perturbations.

The mechanism of coupling is revealed by studying how the coupling between any two sites is affected by the configuration of other sites (Di Cera, 1998a,b). The coupling free energy between two mutations at sites i and j as defined in the double-mutant cycle in eq. (6) applies when all other sites are in state 0. A cycle analogous to that in eq. (6) can be constructed for any configuration of the other $N-2$ sites. There are 2^{N-2} such configurations and $N(N-1)/2$ distinct pairs of sites, leading to a total of $N(N-1)2^{N-3}$ possible thermodynamic cycles and coupling free energies. Not all cycles are independent, because the system only contains $2^N - 1$ independent terms, N of which are site-specific free energies of perturbation ΔG_js, and the remainder are coupling free-energy terms from the second up to the Nth order. Construction of double-mutant cycles cannot generate more information than that contained in the independent coupling terms. Hence, of the $N(N-1)2^{N-3}$ possible cycles, only $2^N - 1 - N$ are necessarily independent (Di Cera, 1998a,b). However, for any pair of sites i and j, the 2^{N-3} coupling free-energy values generated by the configurations of the other $N-2$ sites are all independent. Therefore, there are two alternative and equivalent ways to characterize the interactions of a system. One is based on the second and higher-order interaction free energies that define the configurations of the system, the other casts these free energies in terms of the coupling between two sites in any possible configuration of the other sites.

Once coupling free energies are calculated for all possible configurations of the system, it is possible to decipher the code for site–site interactions using the following property of a thermodynamic cycle, the mathematical proof of which is given elsewhere (Di Cera, 1995).

THEOREM: *If the coupling between two sites is direct and involves only second-order interactions, then the coupling free energy is independent of the configuration of other sites. Otherwise, the coupling is indirect and involves interactions higher than second-order*

To understand the significance of this property, it is useful to consider two key examples of direct and indirect coupling. Direct coupling is peculiar to models of nearest-neighbor interactions, like the Koshland–Nemethy–Filmer model of ligand-binding cooperativity (Koshland et al., 1966). In this model, interactions are all pairwise and second-order. Coupling of higher order is simply the result of additive contributions from second-order coupling terms. No matter how two sites are linked to each other and to the rest of the system, the coupling between them remains energetically the same regardless of the configuration of other sites. This has the nontrivial consequence that, when the coupling between a pair of sites is not affected by a third site, one cannot conclude that the third site is not coupled to the pair. In fact, in any nearest-neighbor model where the third site is coupled to each site in the pair, the state of the third site has no effect on the coupling free energy of the pair (Di Cera, 1995,1998a,b).

Indirect coupling manifests itself in a more obvious manner. The Monod–Wyman–Changeux model of concerted allosteric transitions (Monod et al., 1965) provides an example where interactions involve all sites through a linked global conformational change. In this model, sites are always positively coupled, and the other of coupling changes according to the state of other sites as the protein switches from one state to another. Combination of the Koshland–Nemethyl–Filmer and Monod–Wyman–Changeux models into a more general hybrid model accounts for arbitrarily complex mechanisms of coupling (Di Cera, 1995,1998a).

The mechanism of coupling can be identified from analysis of double-mutant cycles, but requires the availability of a high-dimensional manifold of perturbations where the coupling between two sites can be studied as a function of a relatively large number of configurations of other sites. This poses challenging tasks from an experimental standpoint, because construction and expression of mutants of the third or higher-orders in a protein may be problematic. The analysis based on the properties of the coupling free energy appears to be ideally suited for the site-specific dissection of ligand recognition, when most of the perturbations are introduced in small ligands that bind to the protein. Large libraries of peptides containing all the relevant mutant forms can be constructed with ease and, when combined with perturbations in the protein, generate the complexity necessary to dissect all interactions in the system. An example of how this new and powerful approach based on the principles of site-specific thermodynamics can be implemented in practice is offered in the next section.

2. Site-specific Dissection of Cooperativity in Enzyme Specificity

2.1. Substrate Recognition of Serine Proteases

One of the best characterized classes of enzymes is that including the serine proteases of the chymotrypsin family (Rawlings and Barrett, 1993,1994). These enzymes participate in key physiological functions like digestion, blood coagulation, fibrinolysis, cell and humoral immunity, and embryonic development. Proteases involved in digestive processes, like trypsin, have wide specificity and are also found in organisms as primitive as eubacteria. In contrast, proteases involved in the more specialized functions of blood coagulation, fibrinolysis, and complement have narrow specificity are are found with few exceptions only in vertebrates (Neurath, 1984; Patthy, 1990). Activity and specificity of some proteases involved in blood coagulation and the complement system are controlled allosterically by the binding of Na^+, whereas more ancestral proteases and those involved in fibrinolysis are devoid of such property (Dang and Di Cera, 1996).

Serine proteases share a common fold composes of two six-stranded β-barrels of similar structure that pack together asymmetrically to host at their interface the residues of the catalytic triad H57, D102, and S195. General properties of this fold have been discussed in detail (Lesk and Fordham, 1996). Although they have a common catalytic mechanism (Warshel et al., 1989), these enzymes differ widely in specificity. The classical framework to understand protease specificity takes into account interactions made by the enzyme with the substrate at the level of individual sites (Schechter and Berger, 1967). Residues of the substrate interacting with the enzyme are labeled with a P and a number from 1 to N, starting from the scissile bond and moving to the N-terminus. Residues of the enzyme making contacts with the substrate are called *specificity sites* are labeled with an S. The amino acid at P1 of the substrate makes contacts with the specificity site S1 of the enzyme, P2 contacts S2 and so forth. The P residues of the substrate are contiguous in sequence, whereas this is not required for the S residues of the enzyme. Residues downstream to the scissile bond of the substrate are numbered P1′, P2′, and so forth, and the corresponding specificity sites on the enzyme are S1′, S2′, and so on. The scissile bond is positioned between P1 and P1′. The existence of multiple recognition sites effectively narrows down specificity by reducing the probability that the required sequence is found in other substrates. The longer the consensus sequence interacting with the enzyme, the smaller the probability that it will occur in another potential substrate. This framework is also relevant to the study of other binding phenomena, most notably those involving the recognition of small peptides by molecules of the major histocompatibility complex (Klein & Horejsi, 1997).

The basic questions asked in the study of substrate recognition by an enzyme, or peptide recognition by molecules of the major histocompatibility complex, are: (1) How much binding energy is contributed by each subsite? (2) Are the subsites independent or linked in a cooperative manner? These important questions are relevant to drug design and the analysis of structure–function relations. The thermodynamic theory of site-specific cooperativity developed extensively for ligand-binding processes (Di Cera, 1995) offers tools to answer these questions.

2.2. Thrombin Structure and Function

Among serine proteases, thrombin has been studied in considerable detail (Di Cera et al., 1997; Di Cera, 1998c). Thrombin is capable of two important and opposite roles that are at the basis of the efficiency of blood coagulation. The procoagulant role entails the conversion of fibrinogen into the insoluble fibrin clot, the promotion of platelet aggregation, the stabilization of the ensuing clot by activation of factor XIII and inhibition of fibrinolysis, and the feedback enhancement of its own generation from prothrombin by activation of factors V, VIII, and XI. The anticoagulant role involves the thrombomodulin-assisted conversion of protein C into an active component that cleaves and inactivates factors VIIIa and Va together with protein S, thereby limiting the conversion of prothrombin into thrombin catalyzed by the prothrombinase complex (Mann et al., 1990; Davie et al., 1991). In addition to its primary roles in coagulation, thrombin elicits a variety of important effects on a number of cell lines upon binding to its receptors (Grand et al., 1996; Ishihara et al., 1997).

Sodium ions are required for the optimal conversion of fibrinogen into fibrin monomers, which is catalyzed by the procoagulant fast (Na^+-bound) form with high specificity. The slow (Na^+-free) form of thrombin performs the same task with lower specificity. This form, on the other hand, has higher specificity than the fast form toward protein C (Dang et al., 1995,1997a) and plays an anticoagulant role. As a result of the different affinity of the two allosteric forms, Na^+ is actively exchanged in the transition state upon binding of fibrinogen or protein C. Fibrinogen binds to the fast forms with higher affinity, promotes the slow-to-fast conversion and Na^+ binding. On the other hand, binding of protein C promotes the fast-to-slow conversion and Na^+ release. Hence, Na^+ binding and disso-

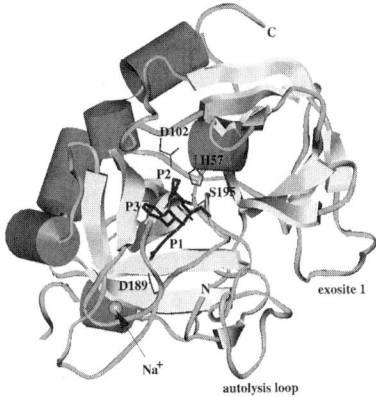

Figure 2.2. Ribbon representation of thrombin showing the residues of the catalytic triad. Important regions of the enzyme are noted.

ciation are important molecular components of substrate recognition by thrombin.

Thrombin is composed of two polypeptide chains of 36 (A chain) and 259 (B chain) residues that are covalently linked through a disulfide bond (Bode et al., 1992). The B chain carries the functional epitopes of the enzyme, and has an overall architecture similar to that of pancreatic serine proteases (Fig. 2.2). The extraordinary specificity of thrombin toward fibrinogen arises not only from contacts made in the interior of the active site (see below), but also from interactions with exosite I located about 2 nm away from the active site (Martin et al., 1992; Stubbs et al., 1992). Another factor that influences thrombin specificity is the W60d insertion loop that is unique to thrombin and defines the apolar specificity site S2. This loop narrows significantly the access to the active site by protruding into the solvent. Replacement of W60d with the less bulky Ala or Ser profoundly affects the interaction of thrombin with the natural inhibitor antithrombin III (Rezaie, 1996), or fibrinogen (Guinto et al., 1995; Guinto and Di Cera, 1997). A similar function has been hypothesized for the autolysis loop shaping the lower rim of the access to the active site. Deletion of the entire loop results in a selective loss of fibrinogen binding (Dang et al., 1997b).

The Na^+-binding site (fig. 2.3) displays octahedral coordination involving the carbonyl O atoms of R221a and K224 and four buried water molecules tetrahedrally coordinated by protein atoms and other water molecules (Di Cera et al., 1995; Zhang and Tulinsky, 1997) that altogether define a complex hydrogen-bonding network within the catalytic pocket (Krem and Di Cera, 1998). Some of the hydrogen bonds in the network are conserved with trypsin (Bartunik et al., 1989). Others are specific to thrombin, and are associated with Na^+ and its coordination shell. The bound Na^+ is located 1.5–2 nm away from the catalytic triad

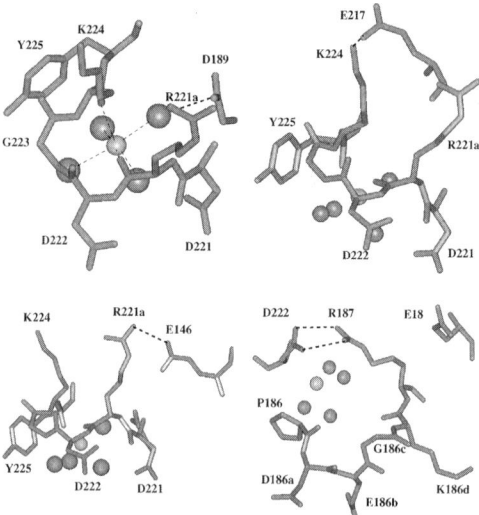

Figure 2.3. Molecular environment of the Na$^+$ binding site of thrombin. The bound Na$^+$ (light circle) is coordinated octahedrally by the carbonyl O atoms of K224 and R221a and four water molecules (dark circles). The site is stabilized by three ion pairs: R221a–E146, D221–R187, D222–R187, and K224–E217.

and lies within 0.5 nm from D189 in the specificity site S1 with a water molecule mediating a hydrogen-bonding interaction with O$^{\delta 2}$ of D189. The Na$^+$ site is stabilized by three ion pairs. R221a is ion-paired to E146 of the autolysis loop, K224 is ion-paired to E217, while D221 and D222 form a bidentate ion pair with R187. Altering the bidentate ion pair with the double substitution D221A/D222K results in reduced activity toward fibrinogen but enhanced activity toward protein C (Di Cera et al., 1995). Perturbation of the same ion pair in the R187Q thrombin Greenville produces a reduced clotting activity, consistent with reduced Na$^+$ binding (Henriksen et al., 1998). Similar effects of reduced clotting activity due to reduced Na$^+$ binding are seen upon disruption of the R221aA-E146 (Miyata et al., 1992; Dang et all., 1997a) or the K224–E217 (Gibbs et al., 1995; Dang et al., 1997a) ion pairs.

2.3. Library of Site-Specific Probes

The molecular strategy used by thrombin to achieve specificity toward fibrinogen and protein C is deeply rooted in the mechanism through which Na$^+$ binding affects the environment of the active site of the enzyme. The main question is how the Na$^+$-induced slow-to-fast conversion enhances specificity toward fibrinogen and small chromogenic substrates. A related question is which allosteric form should be targeted with active-site inhibitors to guarantee optimal specificity. In both cases, the answer resides primarily in the properties of the specificity sites of the enzyme, and warrants a quantitative assessment of their energetic contribution in the transition state.

Substrate libraries generated from combinatorial chemistry or phage display to identify consensus sequences for binding (Babine and Bender, 1997; Smith and Petrenko, 1997) can be used as powerful probes of the molecular environment of the specificity sites of the enzyme to elucidate how they contribute to recognition in the transition state. If perturbations are made in the sequence of a substrate to generate a library containing all species required for a site-specific analysis, much information can be derived on the energetic contributions of the specificity sites that is difficult to obtain from mutagenesis of the enzyme. In order to understand the molecular origin of the higher specificity of the fast form toward fibrinogen, the chromogenic tripeptide substrate FPR (table 2.2) was synthesized (Vindigni et al., 1997) to mimic the interaction of the natural substrate with the active site of the enzyme (Martin et al., 1992; Stubbs et al., 1992). Like fibrinogen, FPR is cleaved by the fast form with a specificity 30-fold higher than that of the slow form (Di Cera et al., 1997; Di Cera, 1998c) (table 2.3). The crystal structure of thrombin inhibited with H-D-Phe-Pro-Arg-CH$_2$Cl (Bode et al., 1992) provides information on the interactions of the P1-3 groups of FPR with the enzyme. Arg at P1 makes an ion pair with D189 at S1 at the bottom of the catalytic pocket, and Pro at P2 interacts with the apolar moiety of S2 defined by P60b, P60c, and W60d, whereas Phe at P3 forms a favorable edge-to-face interaction with the aromatic ring of W215 at S3 (fig. 2.4). The D enantiomer at P3 mimics the interaction of F8 at P9 of fibrinogen with W215 of thrombin (Ni et al., 1992). The chromogenic group paranitroanilide attached to the C-terminus enables quantitative spectroscopic measurements of the released paranitroaniline upon cleavage by thrombin at the P1–paranitroanilide scissile bond.

Table 2.2. Substrate Library

Abbreviation	Substrate	Site(s) perturbed
FPR	H-D-Phe-Pro-Arg-*p*-nitroanilide	none
FPK	H-D-Phe-Pro-Lys-p-nitroanilide	P1
FGR	H-D-Phe-Gly-Arg-p-nitroanilide	P2
VPR	H-D-Val-Pro-Arg-p-nitroanilide	P3
FGK	H-D-Phe-Gly-Lys-p-nitroanilide	P1 and P2
VPK	H-D-Val-Pro-Lys-p-nitroanilide	P1 and P3
VGR	H-D-Val-Gly-Arg.-p-nitroanilide	P2 and P3
VGK	H-D-Val-Gly-Lys-p-nitroanilide	P1,P2, and P3

Table 2.3. Specificity Constants k_{cat}/K_m (in $\mu M^{-1} s^{-1}$) for the Hydrolysis of Synthetic Substrates by Thrombin, tPA, Trypsin, and Plasmin

	FPR	FPK	FGR	VPR	FGK	VPK	VGR	VGK
Thrombin fast form:								
wt	90	7.9	2.0	100	0.021	2.1	0.34	0.0047
R221aA	80	4.6	0.75	36	0.011	0.96	0.14	0.0024
K224A	44	7.7	0.93	24	0.027	1.4	0.17	0.0044
R221aA/K224A	26	3.2	0.33	13	0.011	0.70	0.049	0.0017
Thrombin slow form:								
wt	3.0	0.35	0.86	6.7	0.0026	0.11	0.17	0.00079
R221aA	1.6	0.040	0.042	1.0	0.00038	0.0097	0.0086	0.00013
K224A	0.47	0.034	0.012	0.28	0.00039	0.0063	0.0020	0.00013
R221aA/K224A	0.34	0.010	0.0025	0.077	0.00021	0.0018	0.00063	0.000063
tPA	0.048	0.0070	0.048	0.021	0.00066	0.0039	0.010	0.0013
Trypsin	8.9	0.95	2.2	6.9	0.22	0.75	0.67	0.069
Plasmin	0.031	0.047	0.0018	0.028	0.048	0.058	0.0016	0.0037

Experimental conditions: 5 mM Tris, I=200 mM, 0.1% PEG, pH 8.0 at 25°C. The slow form was studied in the presence of 200 mM choline chloride. The properties of the fast form refer to the limit $[Na^+] \to \infty$, at constant I=200 mM. Errors are typically ±2%.

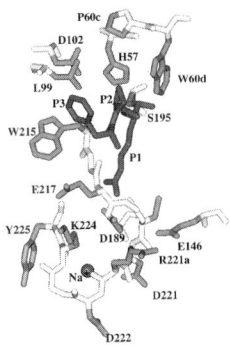

Figure 2.4. Contacts between the irreversible inhibitor H-D-Phe-Pro-Arg-CH$_2$Cl and the active site of thrombin according to the crystal structure (Bode et al., 1992). Shown are thrombin residues D189, P60c, W60d, L99 and W215 that interact with the inhibitor. The guanidyl group of the Arg at P1 makes an ion pair with the carboxyl group of D189 at S1 at the bottom of the active site. Pro at P2 packs in the S2 apolar cavity provided by the 60d loop. The H-D-Phe group at P3 makes favorable hydrophobic contacts in the cleft with L99 and especially a perpendicular aryl–aryl edge-on interaction with W215 at S3.

Starting from FPR, seven substitutions were made to generate the library in table 2.2 (Vindigni et al., 1997). The main idea was to introduce enough perturbation at P1, P2, and P3 while retaining sufficient specificity for accurate experimental measurements. The perturbation would then act as the source of information on the environment of the specificity sites of the enzyme S1, S2, and S3. The group H-D-Phe was replaced by H-D-Val in VPR, VPK, VGR, and VGK, to replace the aromatic moiety with a hydrophobic one present in natural substrates (Di Cera et al., 1997). Pro was replaced by Gly in FGR, FGK, VGR, and VGK, to avoid steric hindrance with S2 and relieve the rigidity of the P2–P3 bond. Arg was replaced by Lys in FPK, FGK, VPK, and VGK, to preserve the positive charge at P1 needed to contact D189 at S1. These substitutions generate all possible intermediates from the parent substrate FPR: the three singly-substituted substrates FPK, FGR, and VPR; the three doubly-substituted substrates FGK, VPK, and VGR; and the triply substituted substrate VGK. The library therefore maps the eight possible intermediates in a three-site system where each site can exist in two states: wild type and mutant.

To obtain the relevant free-energy changes associated with the perturbations, the specificity constant $s = k_{cat}/K_m$ for substrate hydrolysis was

Table 2.4. Free-energy Values (in kcal/mol) Due to Perturbation of the P1–3 Sites of FPR

	ΔG_1	ΔG_2	ΔG_3	ΔG_{12}	ΔG_{13}	ΔG_{23}	ΔG_{123}
Thrombin fast form:							
wt	1.4	2.3	−0.1	1.3	0.8	1.1	2.2
R221aA	1.7	2.8	0.5	0.8	0.5	0.5	1.2
K224A	1.0	2.3	0.4	1.1	0.7	0.6	1.8
R221aA.K224A	1.2	2.6	0.4	0.8	0.5	0.7	1.5
Thrombin slow form:							
wt	1.3	0.7	−0.5	2.2	1.2	1.4	3.3
R221aA	2.2	2.2	0.3	0.6	0.6	0.7	1.0
K224A	1.6	2.2	0.3	0.5	0.7	0.8	0.8
R221aA/K224A	2.1	2.9	0.9	−0.6	0.1	−0.1	−0.8
tPA	2.5	0.0	0.5	−0.0	−0.1	0.4	0.5
Trypsin	1.3	0.8	0.2	0.0	−0.0	0.6	0.6
Plasmin	−0.3	1.7	0.1	−0.3	−0.2	0.0	−0.2

Values were obtained from the specificity constants in table 2.3, using eqs. (7)–(13) in the text. Errors are ±0.1 kcal/mol or less.

measured in all cases (table 2.3) to estimate the free energy of stability of the transition state. The value of FPR was used to scale energetically all others to obtain the free-energy values in table 2.4 as follows (Vindigni et al., 1997; Di Cera, 1998b)

$$\Delta G_1 = -RT \ln(s_{FPK}/s_{FPR}), \quad (7)$$

$$\Delta G_2 = -RT \ln(s_{FGR}/s_{FPR}), \quad (8)$$

$$\Delta G_3 = -RT \ln(s_{VPR}/s_{FPR}), \quad (9)$$

$$\Delta G_{12} = -RT \ln(s_{FGK}s_{FPR}/s_{FPK}s_{FGR}), \quad (10)$$

$$\Delta G_{13} = -RT \ln(s_{VPK}s_{FPR}/s_{FPK}s_{VPR}), \quad (11)$$

$$\Delta G_{23} = -RT \ln(s_{VGR}s_{FPR}/s_{FGR}s_{VPR}), \quad (12)$$

$$\Delta G_{123} = -RT \ln(s_{VGK}s_{FPR}^2/s_{FPK}s_{FGR}s_{VPR}). \quad (13)$$

Here ΔG_1, ΔG_2, and ΔG_3 are the changes in specificity due to the single-site substitutions at P1, P2, and P3; ΔG_{12}, ΔG_{13}, and ΔG_{23} are the second-order coupling free energies for substitutions made at the three possible pairs of sites; and ΔG_{123} is the third-order coupling free energy for the triple substitution. These terms reflect interactions between substitutions made at different sites that may reduce ($\Delta G > 0$) or enhance ($\Delta G < 0$) specificity beyond simple additivity. The terms in eqs. (7)–(13) define the free-energy level of any substrate in the library relative to FPR. The relevant expressions are

$$s_{\text{FPK}} = s_{\text{FPR}} \exp\left(-\frac{\Delta G_1}{RT}\right), \tag{14}$$

$$s_{\text{FPK}} = s_{\text{FPR}} \exp\left(-\frac{\Delta G_2}{RT}\right), \tag{15}$$

$$s_{\text{VPR}} = s_{\text{FPR}} \exp\left(-\frac{\Delta G_3}{RT}\right), \tag{16}$$

$$s_{\text{FGK}} = s_{\text{FPR}} \exp\left(-\frac{\Delta G_1 + \Delta G_2 + \Delta G_{12}}{RT}\right), \tag{17}$$

$$s_{\text{VPK}} = s_{\text{FPR}} \exp\left(-\frac{\Delta G_1 + \Delta G_3 + \Delta G_{13}}{RT}\right), \tag{18}$$

$$s_{\text{VGR}} = s_{\text{FPR}} \exp\left(-\frac{\Delta G_2 + \Delta G_3 + \Delta G_{23}}{RT}\right), \tag{19}$$

$$s_{\text{VGK}} = s_{\text{FPR}} \exp\left(-\frac{\Delta G_1 + \Delta G_2 + \Delta G_3 + \Delta G_{123}}{RT}\right). \tag{20}$$

The same approach was extended to thrombin mutations that perturb Na$^+$ binding and the region around the specificity site S1, in order to pinpoint the molecular signatures of the effects observed in the wild type.

2.4. Cooperativity in Substrate Recognition by Thrombin

Inspection of table 2.4 reveals the presence of large and significant cooperativity in the effects induced by perturbations of the P1–3 sites in the case of wild-type thrombin. The extent of cooperativity changes for each pair of substitutions and is also affected by the allosteric state of the enzyme and mutations made around the Na$^+$ binding environment. In contrast, no interactions are seen for trypsin, plasmin, and tPA, three cognate proteases. The different response elicited by the substrate library in different enzymes lends validity to the strategy of probing the environment of the specificity sites.

The free energy change due to replacing Arg by Lys at P1 in all possible combinations of the state of P2 and P3 is summarized in table 2.5. The values are all positive in both the slow and fast forms, for wild-type and mutant thrombins, indicating that the Arg→Lys replacement at P1 always causes a loss of specificity. The cost of this replacement is about 1 kcal/mol in both the slow and fast forms when no replacement is made at P2 and P3, which suggests that the same mechanism may cause the loss of specificity in both allosteric forms. The changes in catalytic parameters observed in the fast→slow conversion of thrombin for both synthetic substrates and fibrinogen involve a decrease in k_{cat} and an increase in K_{m} (Wells and Di Cera, 1992; Vindigni and Di Cera, 1996). This would suggest that binding of Na$^+$ orients the side chain of D189 for optimal coordination of the guanidinium group of Arg at P1, perhaps using water 447 that bridges the bound Na$^+$ and the O$^{\delta 2}$ atom of D189 (Krem & Di

Table 2.5 Free-energy Change (in kcal/mol) in Specificity Due to Perturbation of the P1–3 Sites of FPR

	Fast form				Slow form				Coupling			
	wt	R221aA	K224A	R221aA/K224A	wt	R221aA	K224A	R221aA/K224A	wt	R221aA	K224A	R221aA/K224A
Replacement at P1 (Arg→Lys)												
FPX	1.4	1.7	1.0	1.2	1.3	2.2	1.6	2.1	0.2	−0.5	−0.5	**−0.8**
FGX	2.7	2.5	2.1	2.0	3.4	2.8	2.0	1.5	**−0.7**	−0.3	0.1	0.5
VPX	2.3	2.1	1.7	1.7	2.4	2.7	2.2	2.2	−0.1	−0.6	−0.6	−0.5
VGX	2.5	2.4	2.2	2.0	3.2	2.5	1.6	1.4	−0.6	−0.1	0.5	0.6
Replacement at P2 (Pro→Gly)												
FXR	2.3	2.8	2.3	2.6	0.7	2.2	2.2	2.9	**1.5**	0.6	0.1	−0.3
FXK	3.5	3.6	3.3	3.4	2.9	2.8	2.6	2.3	0.6	**0.8**	**0.7**	**1.1**
VXR	3.4	3.3	2.9	3.3	2.2	2.8	2.9	2.8	**1.2**	0.5	0.0	0.5
VXK	3.6	3.5	3.4	3.6	2.9	2.6	2.3	2.0	**0.7**	**1.0**	**1.1**	**1.6**
Replacement at P3 (Phe→Val)												
XPR	−0.1	0.5	0.4	0.4	−0.5	0.3	0.3	0.9	0.4	0.2	0.1	−0.5
XPK	0.8	0.9	1.0	0.9	0.7	0.8	1.0	1.0	0.1	0.1	0.0	−0.1
XGR	1.0	1.0	1.0	1.1	1.0	0.9	1.1	0.8	0.1	0.1	−0.1	0.3
XGK	0.9	0.9	1.1	1.1	0.7	0.6	0.7	0.7	0.2	0.3	0.4	0.4

Errors are ±0.1 kcal/mol or less. Values were obtained from the data in table 2.4. The difference between the values for the fast and slow forms gives the coupling between the substitution and the slow-to-fast transition. Positive values are indicative of stabilization of the slow form in the transition state, whereas negative values signal stabilization of the fast form. Values of the coupling of magnitude in excess of RT (0.6 kcal/mol) are in bold type.

Cera, 1998). In this case, however, the loss of specificity with the Arg →Lys substitution at P1 would be more pronounced in the fast form. The similarity of effects seen for the two forms argues against a direct influence of the allosteric switch on the position of the side chain of D189. This conclusion is consistent with the observation that water 447 is also present in trypsin, that does not bind Na^+ (Dang & Di Cera, 1996), where it bridges the $O^{\delta 2}$ atom of D189 to the carbonyl O atom of K224. The origin of the increases specificity of the fast form must therefore reside at other specificity sites.

Due to the strong interactions among the P1–3 sites, the cost of replacing Arg by Lys at P1 depends on the residue at P2 and P3 (table 2.5) and reveals the importance of cooperativity in substrate recognition. With Gly at P2, the cost of the Arg→Lys replacement at P1 increases by 1.3 kcal/mol in the fast form and 2.1 kcal/mol in the slow form, introducing a significant difference of −0.7 kcal/mol between the two forms. This difference measures the coupling between the replacement at P1 and the slow-to-fast transition. A negative value indicates that the replacement promotes the slow-to-fast conversion in the transition state, or that the replaced residues binds preferentially to the slow form. A positive value signals a stabilization of the slow form, or that the replaced residues bind preferentially to the fast form. The presence of a small but significant coupling when Gly is present at P2 suggests that the environment around D189 in the transition state may be different in the slow and fast forms. When P3 is substituted, the energetic penalty for the P1 substitution increases by nearly 1 kcal/mol in both thrombin forms. The extent of interaction of P2 and P3 with P1 is significant. When Gly is present at P2, the interaction with P1 actually exceeds the cost of the replacement at P1 itself in the slow form.

The free energy change due to replacing Pro by Gly at P2 in all possible combinations of the state of P1 and P3 is summarized in table 2.5. As for the substitution at P1, the values are significantly positive. In this case, the effects tend to be more pronounced in the fast form, underscoring an obvious change in the environment of the S2 site in the slow-to-fast transition. The significant difference is conducive to stabilization of the slow form in the transition state when Pro is replaced by Gly. The apolar site S2 of thrombin is formed by residues in the W60d loop that has no counterpart in other serine proteases. Residues in the apolar site must be oriented differently in the slow and fast forms, causing a better discrimination of the residue at P2 in the fast form. The W60d loop may play a key role in this respect, because replacement of the bulky side chain by Ser in W60dS abolishes the differences between the slow and fast forms in recognizing substrates with Pro or Gly at P2 (Guinto and Di Cera, 1997). The indole ring of W60d likely produces steric hindrance in the slow form, but not in the fast form. The perturbation at P2 depends strongly on the residue present at P1 and P3. The cost of the Pro→Gly replacement increases by 1.2 kcal/mol in the fast form and 2.2 kcal/mol in the slow form as a result of the substitution at P1. This effect is exactly (taking into account roundoff error) the same as that seen for the perturbation at P1 when P2 is perturbed, as a consequence of the reciprocity of the linkage between the perturbations at P1 and P2.

The free-energy change due to replacing Phe by Val at P3 in all possible combinations of the state of P1 and P2 is summarized in table 2.5. The Val at P3 increases specificity slightly in the slow form. The hydrophobic group at P3 may interact favorably with the hydrophobic moiety of L99 (Fig. 2.4), which is close to the apolar site S2. Interestingly, residue Y3 of hirudin contacts W215 of thrombin in a manner similar to Phe at P9 of the fibrinogen Aα chain, but replacement of Y3 with more hydrophobic residues significantly enhances the binding affinity (De Filippis et al., 1995,1997), consistent with the enhanced specificity of VPR compared to FPR. The energetic effect linked to replacement of the residue at P3 is of the same magnitude in both forms and excludes a direct involvement of the S3 site in the slow–fast equilibrium. The perturbation at P3 depends strongly on the state of P1 and P2. The cost of the Phe→Val replacement increases by 0.9 kcal/mol in the fast form and 1.2 kcal/mol in the slow form as a result of the substitution at P1, and is the reciprocal of the effect seen for the perturbation at P1 when P3 is perturbed.

The data in tables 2.4 and 2.5 reveal the presence of coupling among perturbations at P1, P2, and P3. The coupling is the result of constraints imposed by the enzyme on the bound substrate in the transition state, and is therefore revealing of the molecular environment underlying the recognition process. The coupling free energies for the three possible pairs of P sites in the two possible states of the third site are listed in table 2.6. The values are constructed from the specificity constants pertaining to the four species generated by substitutions at the P sites. For example, the cou-

Table 2.6. Coupling Free Energies (in kcal/mol) for Perturbation of the P1–3 Sites of FPR

	Fast form				Slow form			
	wt	R221aA	K224A	R221aA/ K224A	wt	R221aA	K224A	R221aA/ K224A
$^0\Delta G_{12}$	1.3	0.8	1.1	0.8	2.2	0.6	0.5	−0.6
$^1\Delta G_{12}$	0.2	0.3	0.6	0.3	0.7	−0.3	−0.6	−0.9
$^0\Delta G_{13}$	0.8	0.5	0.6	0.5	1.2	0.6	0.7	0.1
$^0\Delta G_{13}$	−0.2	−0.1	0.1	−0.0	−0.2	−0.3	−0.4	−0.1
$^0\Delta G_{23}$	1.1	0.5	0.6	0.7	1.4	0.7	0.7	−0.1
$^1\Delta G_{23}$	0.1	−0.0	0.1	0.2	0.0	−0.2	−0.3	−0.3

Listed are the two possible configurations of the third P site (0=wild-type, 1=mutant). Errors are ±0.1 kcal/mol or less.

pling between P1 and P2 is $^0\Delta G_{12} = -RT\ln(s_{FGK}s_{FPR}/s_{FPK}s_{FGR})$ in the absence of perturbation at P3, and $^1\Delta G_{12} = -RT\ln(s_{VGK}s_{VPR}/s_{VPK}s_{VGR})$ when P3 is perturbed. The value of $^0\Delta G_{12}$ is the same as ΔG_{12} in table 2.4. The coupling free energies in the case of wild-type thrombin are mostly positive and quite significant, demonstrating that perturbations at the P1, P2, and P3 sites are negatively coupled in enhancing specificity, and that the residues at P1–3 are negatively coupled in the binding to the S1–3 sites. When a site is perturbed, perturbation at a second site reduces specificity beyond simple additivity. Furthermore, the coupling between any two sites is enhanced by more than 1 kcal/mol when the third site is perturbed, underlying an even stronger cooperative effect in reducing specificity that progresses with the extent of perturbation in the substrate. There are six possible coupling free-energy values for the three pairs, but only four are independent. Hence, the difference between any two values for each pair is exactly the same for all pairs. From the property of the coupling free energy, we conclude that the sites are coupled indirectly through interactions of orders higher than the second.

2.5. Origin of the Higher Specificity of the Fast Form

The Arg→Lys replacement at P1 slightly promotes the slow-to-fast transition when Gly is present at P2. On the other hand, the Pro→Gly replacement at P2 strongly stabilizes the slow form. The replacement at P3 has little effect on the allosteric equilibrium. Hence, the slow-to-fast transition affects mostly the environment of the S2 site, with modest effects on the S1 site and no effect on the S3 site. Constraints at the S2 site accounts for the lower specificity of the slow form compared with the fast form, and become inconsequential if the substrate acquires flexibility with a Gly at P2 and can readjust in the active site to compensate for the increased steric hindrance of the S2 site in the slow form. These findings explain why the thrombin mutant W60dS cleaves FPR with the same specificity in the slow and fast forms (Guinto and Di Cera, 1997) and suggest the bulky side chain of W60d as the likely origin of the constraints at S2.

The dominant factors that control specificity are the rigidity of the P2–P3 bond and the strength of the P1–S1 interaction. When the P2–P3 bond is rigid, the substrate finds a more favorable S2 environment in the fast form. Flexibility of the P2–P3 bond relaxes the optimal interaction of Arg at P1 with D189 at S1, this effect being favored by a more accessible active site in the fast form (Wells & Di Cera, 1992; Ayala & Di Cera, 1994). The coupling between substitutions at P1 and P2 comes partially from an intrinsic effect on the substrate—the loss of rigidity of the P2–P3 bond—and partially from the different environment of the enzyme in the slow and fast forms. The less constrained environment of the specificity sites in the fast form also act to reduce the extent of negative coupling among the various perturbations in the substrate, causing the interactions to essentially disappear as more substitutions are introduced at the P sites.

The two ion pairs R221a–146 and K224–E217 stabilizing the Na^+ binding environment (Fig. 2.3) provide other constraints in the slow form. The R221aA mutant has a reduced Na^+ affinity (Dang et al., 1997a), suggesting that disruption of the R221aA–E146 ion pair may destabilize the fast form. However, disruption of the R221aA–E146 ion pair affects specificity more in the slow than the fast form. The parameters pertaining to the fast form are practically unchanged relative to the wild type, while those in the slow form show enhanced sensitivity to perturbation at P1 and P2. This perturbation is also less dependent on the state of other groups, indicating a reduction in the coupling among substitutions at the P1–3 sites (tables 2.4 and 2.5).

Disruption of the R221a–E146 ion pair has a direct influence on the specificity sites S1 and S2 of the enzyme in the slow form, and affects the way these sites discriminate between Arg and Lys at P1 or Pro and Gly at P2. This ion pair maintains the correct architecture of the S1 site, especially in the slow form, but also influences the S2 site located some 1.7 nm away. The molecular basis of this effect may be due to enhanced mobility of the autolysis loop on the Glu side of the ion pair upon disruption of the contact. The enhanced mobility may interfere with substrate recognition in the slow form. The R221a–E146 ion pair contributes to the integrity of the S1 environment in the slow form—but not in the fast form, because the perturbation is practically abolished by Na^+ building.

As for the R221aA mutant, mutation of K224 to Ala reduces the Na^+ affinity (Dang et al., 1997a) suggesting that disruption of the K224–E217 ion pair may destabilize the fast form, but again this proposal is contradicted by the experimental data that document a larger perturbation of the slow form (tables 2.4 and 2.5). Disruption of the K224–E217 ion pair produces effects very similar to those seen for the R221aA mutant, with a reduction of the coupling among the P1–3 sites especially in the slow form. The ion pair between K224 and E217 bridges two residues on the last two β-strands of the B chain, and contributes to the integrity of the S1 and S2 environments in the slow form. The region in immediate proximity to K224 and E217 plays a key role in substrate selectivity, and is absolutely conserved in thrombin from different species (Banfield & MacGillivray, 1992). The state of this ion pair can therefore control the access of substrates into the bottom of the catalytic pocket where the specificity site S1 is located.

The two ion pairs interact slightly in the slow form, but not in the fast form, as demonstrated by the results on the double mutant R221aA/K224A (tables 2.4 and 2.5). The perturbation induced by the double mutation is more drastic and almost abolishes Na^+ binding (Dang et al., 1997a). The mutation affects the response to perturbations at the P1–3 sites, with an effect more pronounced in the slow form. The site-specific parameters are profoundly altered in the slow form and, interestingly, the pairwise coupling pattern shows the disappearance of indirect coupling in both the slow and fast forms, with the onset of positive second-order direct coupling between P1 and P2 (table 2.6). This effect is peculiar to the double substitution, though it is somewhat anticipated by the single substitutions. The molecular basis for the synergism between the R221a–E146 and K224–E217 ion pairs in the slow form is in the participation of residues R221a and K224 in Na^+ and water coordination. In the fast form, the carbonyl O atoms of R221a and K224 directly ligate the Na^+. Mutation of these residues reduces the Na^+ affinity, but high concentrations of Na^+ oppose the structural perturbation induced by the mutation restoring a molecular environment for the specificity sites that is essentially that of the fast form of wild type. When Na^+ is released, the carbonyl O atom of K224 may reorient, as seen in the structure of trypsin, and may hydrogen-bond to water 447 in concert with the carbonyl O atom of R221a. Water 447 hydrogen-bonds to the side chain of D189 in the specificity pocket S1 and through the switching mechanism any perturbation of R221a and K224 changing the orientation of the carbonyl O atoms will not be compensated as in the case of the fast form and therefore may lead to more drastic structural changes (Krem and Di Cera, 1998).

We conclude that the more constrained environment in the slow form of thrombin is partially due to stronger ion pairs formed by R221a and K224 in the Na^+-binding loop with E146 in the autolysis loop and E217 in the penultimate β-strand of the B chain. The integrity of these ion pairs is essential for maintaining the correct architecture of the specificity sites through the effect on the water molecules in the channel that embeds the specificity site S1. The role of the ion pairs in the fast form appears to be less critical, and their disruption can be compensated by the binding of Na^+

2.6. Molecular Origin of the Cooperativity among the P1–3 Sites

The coupling pattern emerged from the analysis of the substrate library (table 2.6) is conducive to negatively cooperative iteractions higher than second-order. To elucidate the origin of this coupling, derived from the property of the coupling free energy, the entire five-dimensional manifold of species should be considered. This manifold is composed of the sites P1, P2, P3, R221a, and K224, and the relevant free energies are calculated by operating on the values listed in table 2.4. Analysis of the coupling pattern involving all possible pairs (table 2.7) shows how interactions change with the state of other sites. Considering only differences of magnitude at least RT (0.6 kcal/mol) in the coupling free energy, the patterns can be analyzed to identify the nature of the interaction.

Table 2.7. Coupling Free Energies (in kcal/mol) for Perturbation of the P1–3 Sites of FPR and Residues R221a and K224 of Thrombin

	000	100	010	001	110	101	011	111	Coupling	Mediated by
Fast form:										
P1–P2	1.3	0.2	0.8	1.1	0.3	0.5	0.8	0.3	indirect	P3
P1–P3	0.8	−0.2	0.5	0.6	−0.1	0.1	0.5	−0.0	indirect	P2
P1–R221a	0.2	−0.2	−0.1	0.2	−0.1	−0.1	0.0	−0.2	none	
P1–K224	−0.4	−0.6	−0.6	−0.4	−0.4	−0.5	−0.4	−0.4	none	
P2–P3	1.1	0.1	0.5	0.6	−0.0	0.1	0.7	0.2	indirect	P1
P2–R221a	0.5	0.1	−0.1	0.3	−0.1	0.0	0.4	0.1	none	
P2–K224	0.0	−0.2	−0.4	−0.2	−0.2	−0.2	0.0	0.0	none	
P3–R221a	0.2	0.1	−0.1	0.0	0.0	−0.1	0.1	0.0	none	
P3–K224	0.4	0.2	−0.0	−0.1	0.2	−0.0	0.1	0.2	none	
R221a–K224	0.2	0.2	0.0	−0.2	0.1	−0.0	0.2	0.2	none	
Slow form:										
P1–P2	2.2	0.7	0.6	0.5	−0.3	−0.6	−0.6	−0.9	indirect	P3,R221a,K224
P1–P3	1.2	−0.2	0.6	0.7	−0.3	−0.4	0.1	−0.1	indirect	P2
P1–R221a	0.9	−0.6	0.3	0.5	−0.7	−0.6	−0.0	−0.3	indirect	P2
P1–K224	0.3	−1.4	−0.2	−0.1	−1.6	−1.3	−0.5	1.1	indirect	
P2–P3	1.4	0.0	0.7	0.7	−0.2	−0.3	−0.1	−0.3	indirect	P1,R221a,K224
P2–R221a	1.4	−0.1	0.6	0.7	−0.4	−0.4	−0.1	−0.3	indirect	P1,P3,K224
P2–K224	1.4	−0.3	0.7	0.7	−0.6	−0.5	0.0	−0.6	indirect	P1,P3,R221a
P3–R221a	0.7	0.1	−0.0	0.6	−0.1	0.0	−0.2	0.1	indirect	P2
P3–K224	0.8	0.3	0.1	0.6	−0.0	0.2	−0.1	0.1	indirect	P2
R221a–K224	−0.2	−0.6	−0.9	−0.4	−0.8	−0.7	−1.1	−0.6	indirect	P2

Listed are all possible configurations of the other sites (0=wild type, 1-mutant) in the order P1, P2, P3, R221a, K224. Errors are ±0.1 kcal/mol or less. Indirect coupling requires values that differ by at least RT(0.6 kcal/mol). Direct coupling of magnitude less than RT on the average is considered zero.

In the fast form, only the P1–3 sites are significantly coupled and in an indirect way. Perturbation of any P site influences the coupling at other sites. In the slow form, all sites are strongly coupled. Each coupling can be dissected to identify the element perturbing the interaction. A direct way to illustrate the effect of a third site on the coupling between two sites is to calculate the difference in coupling free energy of a pair due to the 0→1 transition of a third site, in all possible configurations of the remaining sites. The P2 site emerges as a major node of interaction. In the slow form, the state of P2 influences all interactions (table 2.7). The state of R221a and K224 influences the P1–P2 and P2–P3 interactions, but has no effect on the P1–P3 coupling that is influenced by P2. Finally, the coupling between R221a and K224 is influenced by P2 only. As a result, the Ala replacements at these thrombin residues produce additive effects on specificity when Pro is at P2, but are positively linked when Pro is replaced by Gly.

It is of interest to note that the molecular determinants of cooperativity among the specificity sites in thrombin, like the region around W60d in the S2 site and the R221a–E146 and K224–E217 ion pairs, are not present in trypsin, plasmin, and tPA. These proteases, unlike thrombin, show simple additivity of the effects of perturbing individual sites in the substrate (table 2.4). Disruption of the R221a–E146 and K224–E217 ion pairs in thrombin produces a trypsin-like energetic profile. The coupling free energies reflect the strain imposed by the enzyme on the substrate in the transition state. More constrained environments, like thrombin in the slow form, tend to couple more the substitutions made at different P sites. In more relaxed environments, like thrombin in the fast form or trypsin, the coupling is greatly reduced or absent.

3. Predicting the Energetics of Thrombin–Substrate Interactions

The detailed characterization of thrombin–substrate interactions based on the principle of site-specific thermodynamics provides a large body of information that can assist in the important task of predicting energetics from structure. Thermodynamics plays a key role also in this aspect of protein energetics. In principle, the free energy involved in a protein–ligand interaction should be amenable of prediction from computational methods if the relevant structural information is available and the correct force field is used. Current approaches to this problem are of two kinds. One is more empirical and attempts to define the free-energy function as a three-component term including enthalpy, conformational entropy, and hydration free energy (Ajay & Murcko, 1995; Vajda et al., 1997). The other is based on integration of the potential energy along the trajectory between starting and final states (Tidor & Karplus, 1994). The former method is more approximate, but also more advantageous from a computational standpoint, and was chosen in our analysis to verify to what extent the energetics of thrombin–substrate interactions and the coupling among substitutions made at the P1–3 sites could be predicted from knowledge of the structure of the complex. The same approach was used in the case of trypsin–substrate interactions where the energetics of the substitutions at the P1–3 positions are simply additive.

The coordinates of thrombin were obtained from the 0.19 nm resolution structure of thrombin–PPACK (Bode et al., 1992) available as 1PPB in the Protein Data Bank, with water 342 correctly assigned as Na^+. A basic assumption was made that the thrombin–substrate complex for any substrate would not deviate from the thrombin–PPACK model. Other models were also tested: the 0.22 nm resolution structure of thrombin bound to the fragment 7–19 of fibrinopeptide A (Martin et al., 1996), the 0.23 nm resolution structure of thrombin bound to the fragment 7–16 of fibrinopeptide A (Martin et al., 1996), the 0.23 nm resolution structure of thrombin bound to the fragment 7–16 of fibrinopeptide A (Martin et al., 1996), the 0.23 nm resolution structure of thrombin bound to hirugen (Zhang & Tulinsky, 1997), available as 1UCY, 1BBR, and 1HAH in the Protein Data Bank. The thrombin–PPACK model gave the best correlation between calculated and experimentally determined values of the free energy of stabilization of the transition state (see below). For this reason, the thrombin–PPACK model was used as reference for all other thrombin models. The structure 1TLD (Bartunik et al., 1989) was used for trypsin. Hydrogen atoms were added to the structures at pH 7.0 using BIOPOLYMER (MSI, San Diego, CA). Amino-acid substitutions and chemical derivatization were also performed with BIOPOLYMER. The various thrombin–substrate complexes in the library in table 2.2 were built by substituting the desired group at the P1, P2, and P3 positions. Three ligation states were tested for each substrate: (a) the *substrate* state with the intact chromogenic group not covalently bound to the enzyme; (2) the *product* state without the chromogenic group and with the P1–3 groups not covalently bound to the enzyme; (3) the *intermediate* state where the P1–3 tripeptide is covalently bound to His57 and Ser195 of the enzyme as in the thrombin–PPACK model (Bode et al., 1992).

Three sets of coordinates were built for each thrombin–substrate complex. The first had all water molecules removed (dry system). The second had all the water molecules reported in the original crystal structure (wet system). The third had water molecules added to the original ones in the crystal structure to fill a $6 \times 6 \times 6$ nm^3 cube with periodic boundary conditions (soaked system). Three protocols were applied in each case: minimization, with 300 steepest-descent and then 3000 conjugated-gradient steps; Monte Carlo, using rotation of ligand and binding-site torsional angles during heating and cooling cycles; and molecular dynamics, by simulated annealing during 50 ps. Simulations were tested with Discover

2.9 (MSI, San Diego, CA), Xplor 3.6 (Brunger et al., 1986), and AMBER 3.0 (Weiner et al., 1986) with different force fields and parameter sets. Final conformations were checked with Procheck 3.0 (Laskowski et al., 1993). Only computer-generated models of thrombin–substrate complexes with an r.m.s. deviation below 0.15 nm for the thrombin backbone and under 0.2 nm for FPR heavy atoms compared to the starting conformations were considered acceptable. In no case was a good model produced when coordinates were stripped from water molecules (dry model). In the wet and soaked coordinates at least one model per simulation was found to satisfy the constraints on the r.m.s. deviation for thrombin and the substrate.

The best correlation between calculated and experimental values for the binding free energy was obtained with the wet models of tripeptides covalently bound to thrombin (figure 2.5), minimized with Discover using the CFF91 force field, a unit dielectric constant ε, and no distance cutoff for the unbound atom pairs. All thrombin conformations were close to the original X-ray data, and the substrate conformations were within 0.1 nm in the backbone and 0.3 nm in the side chains compared to the original model 1PPB (Bode et al., 1992).

The predicted free energy of association between thrombin (T) and substrate (S) in the transition state, ΔG, was calculated from the thermodynamic cycle

$$
\begin{array}{ccccccc}
 & & & \Delta G_{gas} & & & \\
\text{gas} & S & + & T & \Rightarrow & TS & \\
 & \Delta G_{S,hyd} \Downarrow & \Delta G_{T,hyd} \Downarrow & & & \Downarrow \Delta G_{TS,hyd} & \quad (21) \\
\text{sol} & S & + & T & \Rightarrow & TS & \\
 & & & \Delta G & & &
\end{array}
$$

as

$$\Delta G = \Delta G_{gas} + \Delta G_{hyd} \qquad (22)$$

with

$$\Delta G_{hyd} = \Delta G_{TS,hyd} - \Delta G_{T,hyd} - \Delta G_{S,hyd}. \qquad (23)$$

The value of ΔG_{gas} was calculated from its enthalpic and entropic contributions expressed as

$$\Delta H_{gas} = E_{TS,vdw} + E_{TS,coul} - E_{T,vdw} - E_{T,coul} - E_{S,vdw} - E_{S,coul}, \qquad (24)$$

$$\Delta S_{gas} = \Delta S_{rt,gas} + \Delta S_{conf,gas} + \Delta S_{vib,gas}. \qquad (25)$$

The enthalpy ΔH_{gas} is a function of the van der Waals and coulombic components, whereas ΔS_{gas} is defined in terms of the rotational, configurational, and vibrational components. The van der Waals and coulombic components were computed from the CFF91 force field with $\varepsilon = 2$, without distance dependence or cutoff.

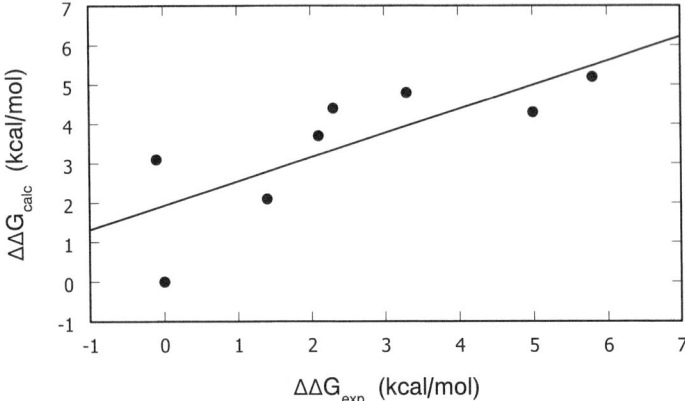

Figure 2.5. Correlation plot between the experimentally determined values of $\Delta G = RT \ln s$ relative to that of FPR and those calculated from analysis of the crystal structure of the fast form (Bode et al, 1992). The correlation factor is 0.77 (see Table 2.8). The values of the predicted $\Delta\Delta G$ in the ordinate are given in table 2.8.

The value of $\Delta S_{\text{rt,gas}}$ was fixed as reported by Finkelstein and Janin (1989). This value depends on the logarithm of the size of the molecule. The size of the different substrates are very similar, and the resulting difference in $\Delta S_{\text{rt,gas}}$ is vanishingly small. The value of $\Delta S_{\text{conf,gas}}$ was computed from the empirical scale of side-chain and main-chain rotation freedom according to Pickett and Sternberg (1993) using the definition

$$T\Delta S_{\text{conf,gas}} = \sum_i \Delta f(r_i) T \Delta s_i + RT \ln\left(\frac{\Delta[\sigma(\phi_i)\sigma(\psi_i)]}{360 \times 360 \rho_i}\right), \qquad (26)$$

where $f(r_i) = r_i^8/(r_i^8 + 500)$ and r_i is the relative accessibility of the ith residue side chain. Δs_i values were taken from Pickett and Sternberg (1993). The standard deviations of main-chain dihedrals, $\sigma(\phi_i)$ and $\sigma(\psi_i)$ were taken from a 10 ps dynamic-trajectory analysis of complexes and free molecular components at 300 K. The accessible-area fraction for residue dihedral pairs in the Ramachandran graph, ρ_i, was fixed at 0.28 for Pro, 0.56 for Gly, and 0.40 for all other amino acids according to allowed and core region for dihedral pairs in Procheck graphs.

The entropy change $\Delta S_{\text{vib,gas}}$ was computed using Discover only for FPR, thrombin, and the thrombin-FPR complex. The different species were carefully minimized (5000 GC cycles) before normal-mode analysis from dynamic simulations. In all models, the side chains of the peptides occupy the same thrombin pocket and interact with the same residues, lending support to the hypothesis that the low-frequency motion of the complexes is similar and that the value of $\Delta S_{\text{vib,gas}}$ is the same for all complexes.

The values of $\Delta G_{S,hyd}$, $\Delta G_{T,hyd}$, were computed using five different methods (Eisenberg & McLachlan, 1986; Ooi et al., 1987; Nicholls & Honig, 1991; Vila et al., 1991; Wesson & Eisenberg, 1992). The best results were obtained by applying the method of Ooi et al. (1987) to our molecular system.

Based on the above definitions, the free energy for the thrombin–FPR complex becomes

$$\Delta G(\text{FPR}) = \Delta H_{gas}(\text{FPR}) - T\Delta S_{rt,gas}(\text{FPR}) - T\Delta S_{vib,gas}(\text{FPR}) \\ - T\Delta S_{conf,gas}(\text{FPR}) + \Delta G_{hyd}(\text{FPR}). \quad (27)$$

The free energy of the XXX substrate relative to FPR is then

$$\Delta\Delta G(\text{XXX}) = \Delta G(\text{XXX}) - \Delta G(\text{FPR}) = \Delta H_{gas}(\text{XXX}) - TS_{conf,gas}(\text{XXX}) \\ + \Delta G_{hyd}(\text{XXX}) - \Delta H_{gas}(\text{FPR}) - T\Delta S_{conf,gas}(\text{XXX}) + \Delta G_{hyd}(\text{XXX}). \quad (28)$$

The predicted values of free energy were all expressed relative to that of FPR and are given in table 2.8 for a direct comparison with the experimental data reported in table 2.4. A plot showing the correlation between calculated and experimental values is given in fig. 2.5. The calculated values in this plot refer to the 1PPB model, with the crystallographic water molecules included and with the substrate bound in the *intermediate* state as in the PPACK model. Removal of Na^+ from the structure yielded a correlation factor with the data pertaining to the slow form of 0.76. The same approach applied to trypsin gave the results summarized in table 2.9. The good correlation obtained in this case where the energetics of substrate recognition at the S1–3 sites are additive lends support to the validity of our approach.

Table 2.8. Calculated Energy Balance (in kcal/mol) due to Perturbation of the P1–3 Sites of FPR on Thrombin

	FPR	FPK	FGR	VPR	FGK	VPK	VGR	VGK	r
$\Delta\Delta G$	0.0	2.1	4.4	3.1	4.3	3.7	4.8	5.2	0.77
$\Delta\Delta E_{vdw}$	0.0	0.9	0.1	−0.1	2.5	1.9	1.5	3.3	0.91
$\Delta\Delta E_{coul}$	0.0	3.1	3.3	4.0	4.6	5.8	5.2	6.9	0.72
$-T\Delta\Delta S_{conf,gas}$	0.0	0.0	0.1	0.0	0.0	−0.1	−0.1	−0.1	0.40
$\Delta\Delta G_{hyd}$	0.0	−1.0	1.0	−0.9	−0.3	−1.9	−0.4	−1.6	0.20

Values of the free energy for substrate binding in the transition state, relative to FPR, were calculated using the atomic coordinates of the crystal structure 1PPB (Bode et al., 1992). Energetic terms are defined in eqs. (21)–(28) in the text. Here r is the correlation factor of each energetic term with the experimental data in table 2.4, relative to FPR, for thrombin in the fast form (see also fig. 2.5).

Table 2.9. Calculated Energy Balance (in kcal/mol) Due to Perturbation of the P1–3 Sites of FPR on Trypsin

	FPR	FPK	FGR	VPR	FGK	VPK	VGR	VGK	r
$\Delta\Delta G_{exp}$	0.0	1.3	0.8	0.2	2.2	1.5	1.5	2.9	–
$\Delta\Delta G$	0.0	0.7	1.2	1.2	2.0	1.5	1.3	3.2	0.87
$\Delta\Delta E_{vdw}$	0.0	0.5	0.0	0.3	0.8	1.0	0.6	1.8	0.88
$\Delta\Delta E_{coul}$	0.0	1.4	1.3	1.0	2.1	1.7	0.8	2.5	0.87
$-T\Delta\Delta S_{conf,gas}$	0.0	−0.2	−0.1	0.0	0.0	−0.1	0.0	−0.1	0.39
$\Delta\Delta G_{hyd}$	0.0	−1.0	−0.1	−0.1	−0.9	−1.1	−0.1	−1.0	0.78

Values of the free energy for substrate binding in the transition state, relative to FPR, were calculated using the atomic coordinates of the crystal structure 1TLD (Bartunik et al., 1989). Energetic terms are defined in eqs. (21)–(28) in the text. Here r is the correlation factor of each energetic term with the experimental value, ΔG_{exp}, relative to FPR, for trypsin.

The dissection of the free-energy values for the various tripeptide substrates of thrombin in the library is given in table 2.8. Enthalpy emerges as the major energetic component in the recognition process. The van der Waals potential of thrombin–substrate interaction correlates strongly with the experimentally determined values, probably because the polarity of the substrate is not significantly perturbed in the P1–3 substitutions. The determining factor seems to be the decrease of the side-chain volumes relative to FPR that displays an optimal fit in the specificity sites S1–3.

Figure 2.6a shows a comparison of the conformations of the bound substrates with Arg (black) or Lys (grey) at P1. The shorter side chain of Lys increases the distance with the carboxylates of D189 at the bottom of the S1 pocket and produces a loss of affinity. The amine position deviates up to 0.16 nm. Figure 2.6b shows a comparison of the conformations of the bound substrates with Pro (black) or Gly (grey) at P2. The main effect of the Pro→Gly substitution is to reduce the volume of the substrate—by 0.06 nm^3—providing increased mobility within the binding pocket. The loss of van der Waals contacts in the Gly derivatives, along with the increased mobility of the P2–P3 bond illustrated by the rotation of the phenyl ring of P3 up to 20°, may bring about the reduced affinity observed experimentally and predicted by calculations. Figure 2.6c shows a comparison of the conformations of the bound substrates with Phe (black) or Val (grey) at P3. As for the substitution at P2, the Phe→Val substitution reduces the volume of the substrate, by 0.054 nm^3, although good van der Waals contacts are retained in the Val derivatives. This explains the smaller loss of affinity observed in these derivatives relative to FPR. It is noteworthy that the substitution at P3 does not seem to affect the orientation of residues at P2 and P1, consistent with the very small coupling between P3 and the other two sites observed experimentally. These results are consistent with the conclusions drawn from analysis of experimental data in section 2.

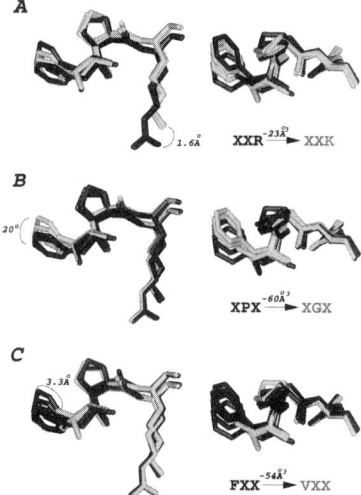

Figure 2.6. Overlay of peptides from their respective peptide-thrombin complex. The eight complexes are superimposed according to the best RMSD of thrombin backbones. Left and right views are rotated 90°. (a) Peptides with Arg (black) or Lys (grey) at P1. (b) Peptides with Pro (black) or Gly (grey) at P2. (c) Peptides with Phe (black) or Val (grey) at P3. The volume change due to the amino acid substitution was calculated as described by Harpaz et al. (1994) and is indicated in each case. Also shown are the main geometrical deviations.

4. Conclusions

The approach to enzyme–substrate energetics based on site-specific thermodynamics is capable of unraveling the extent and origin of coupling among specificity sites responsible for recognition. This novel approach provides information on enzyme–substrate interactions that can be used to optimize the synthesis of active-site inhibitors or more specific substrates. Substrate libraries constructed with proper substitutions at the P positions effectively probe the environment of the specificity sites of the enzyme and detect how they are perturbed upon site-directed mutations or the binding of allosteric effectors. Much will be learned from a systematic application of this approach to other enzyme systems and their chemically modified derivatives.

An important conclusion derived from the analysis of Ala-scanning mutagenesis of protein epitopes and thrombin–substrate interactions is that cooperativity plays a crucial role in ligand recognition. Computational derivation of thermodynamic quantities like the free energy of transition-state stabilization indicates the the cooperative nature of the recognition process can be predicted with reasonable accuracy.

Relative changes in the binding free energy calculated for a given substrate are consistent with those derived experimentally. The good correlation facilitates the prediction of binding free energies for substrates and inhibitors to be synthesized, and may reveal aspects of the enzyme-substrate interaction that are difficult to anticipate from a small set of ligands. The interplay between experimental and computational approaches guided by thermodynamic principles will provide new and important clues to the origin of biological specificity and to ways of manipulating it for practical purposes.

Acknowledgments This work was supported by NIH Research Grants HL49413 and HL:58141. Thierry Rose is the recipient of a fellowship from the Institut Pasteur, NATO and SANOFI-Thrombose Association.

References

Ackers, G. K., and Smith, F. R. (1985) *Ann. Rev. Biochem.* **54**, 597–629.
Ajay, A., and Murcko, M. A. (1995) *J. Med. Chem.* **38**, 4953–67.
Ayala, Y. M., and Di Cera, E. (1994) *J. Mol. Biol.* **235**, 733–46.
Babine, R. E., and Bender, S. L. (1997) *Chem. Rev.* **97**, 1359–472
Banfield, D. K., and MacGillivray, R. T. A. (1992) *Proc. Natl. Acad. Sci. USA.* **89**, 2779–83
Bartunik, H. D., Summers, L. J., & Bartsch, H. H. (1989) *J. Mol. Biol.* **210**, 813–28.
Bode, W., Turk, D., and Karshikov, A. (1992) *Protein Sci.* **1**, 427–71.
Brunger, A. T., Clore, G. M., Gronenborn, A. M., and Karplus, M. (1986) *Proc. Natl. Acad. Sci. USA* **83**, 3801–5.
Carter, P., and Wells, J. A. (1988) *Nature* **332**, 564–8.
Carter, P. J., Winter, G., Wilkinson, A. J., and Fersht, A. R. (1984) *Cell* **38**, 835–40.
Castro, M. J. M., and Anderson, A. (1996) *Biochemistry* **35**, 11435–46.
Clackson, T., and Wells, J. A. (1995) *Science* **267**, 383–6.
Cosmatos, A., Cheng, K., Okada, Y., and Katsoyannis, P. G. (1978) *J. Biol. Chem.* **253**, 6586–90.
Creighton, T. E. (1992) *Protein Folding.* New York, NY: Freeman.
Cunningham, B. C., and Wells, J. A. (1989) *Science* **244**, 1081–5.
Dang, Q. D., and Di Cera, E. (1996) *Proc. Natl. Acad. Sci. USA* **93**, 10253–6.
Dang, Q. D., Guinto, E. R., and Di Cera, E. (1997a) *Nature Biotechnol.* **15**, 146–9.
Dang, Q. D., Sabetta, M., and Di Cera, E. (1997b) *J. Biol. Chem.* **272**, 19649–51.

Dang, Q. D., Vindigni, A., and Di Cera, E. (1995) *Proc. Natl. Acad. Sci. USA* **92**, 5977–81.

Davie, E. W., Fujikawa, K., and Kisiel, W. (1991) *Biochemistry* **30**, 10363–370.

De Filippis, V., Quarzago, D., Vindigni, A., Di Cera, E., and Fontana, A. (1997) *Biochemistry* **37**, 13507–515.

De Filippis, V., Vindigni, A., Alltichieri, L., and Fontana, A. (1995) *Biochemistry* **34**, 9552–64.

Di Cera, E. (1995) *Thermodynamic Theory of Site-Specific Binding Processes in Biological Macromolecules.* Cambridge, U.K.: Cambridge University Press.

Di Cera, E. (1998a) *Adv. Protein Chem.* **51**, 59–119.

Di Cera, E. (1998b) *Chem. Rev.* **98**, 1563–92.

Di Cera, E. (1998c) *Trends Cardiovasc. Med.* **8**, 340–50.

Di Cera, E., Dang, Q. D., and Ayala, Y. M. (1997) *Cell. Mol. Life Sci.* **53**, 701–30.

Di Cera, E., Guinto, E. R., Vindigni, A., Dang, Q. D., Ayala, Y. M., Wuyi, M., and Tulinsky, A. (1995) *J. Biol. Chem.* **270**, 22089–92.

Dickinson, C. D., Kelly, C. R., and Ruf, W. (1996) *Proc. Natl. Acad. Sci. USA* **93**, 14379–84.

Dill, K. A. (1997) *J. Biol. Chem.* **272**, 701–4.

Eisenberg, D., and McLachlan, A. D. (1986) *Nature* **319**, 199–203.

Fersht, A. R., and Serrano, L. (1993) *Curr. Opin. Struct. Biol.* **3**, 75–83.

Finkelstein, A. V., and Janin, J. (1989) *Protein Eng.* **3**, 1–3.

Gibbs, C. S., Coutre, S. E., Tsiang, M., Li, W-X., Jain, A. K., Dunn, K. E., Law, V. S., Mao, C. T., Matsumura, S. Y., Mejza, S. J., Paborsky, L. R., and Leung, L. L. K. (1995) *Nature* **378**, 413–6.

Grand, R. J. A., Turnell, A. S., and Grabham, P. W. (1996) *Biochem. J.* **313**, 353–68.

Green, S. M., Meeker, A. K., and Shortle, D. (1992) *Biochemistry* **31**, 5717–28.

Green, S. M., and Shortle, D. (1993) *Biochemistry* **32**, 10131–9.

Greenspan, N. P., and Di Cera, E. (1999) *Nature Biotechnol.*, **17**, 936–7.

Guinto, E. R. and Di Cera, E. (1997) *Biophys. Chem.* **64**, 103–9.

Guinto, E. R., Vindigni, A., Ayala, Y., Dang, Q. D., and Di Cera, E. (1995) *Proc. Natl. Acad. Sci. USA* **92**, 11185–9.

Guinto, E. R., Caccia, S., Rose, T., Fütterer, K., Waksman, G., and Di Cera, E. (1999) *Proc. Natl. Acad. Sci. USA* **96**, 1852–7.

Harpaz, Y., Gersten, M., and Chotia, C. (1994) *Structure* **2**, 611–49.

Henriksen, R. A., Dunham, C. K., Miller, L. D., Casey, J. T., Menke, J. B., Knupp, C. L., and Usala, S. J. (1998) *Blood* **91**, 2026–31.

Horovitz, A., and Fersht, A. R. (1990) *J. Mol. Biol.* **214**, 613–7.

Horovitz, A., and Fersht, A. R. (1992) *J. Mol. Biol.* **224**, 733–40.

Howell, E. E., Booth, C., Farnum, M., Kraut, J., and Warren, M. S. (1990) *Biochemistry* **29**, 8561–8.

Ishihara, H., Connolly, A. J., Zeng, D., Kahn, M. L., Zheng, Y. W., Timmons, C., Tram, T., and Coughlin, S. R. (1997) *Nature* **386**, 502–6.
Jackson, S. E., and Fersht, A. R. (1993) *Biochemistry* **32**, 13909–18.
Janin, J. (1995) *Prog. Biophys. Molec. Biol.* **64**, 145–66.
Klein, J., and Horejsi, V. (1997) *Immunology*. Oxford, U.K.: Blackwell Science.
Kobayashi, M., Ohgaku, S., Iwasaki, M., Maegawa, H., Watanabe, N., Takada, Y., Shigeta, Y., and Inouye, K. (1984) *Biomed. Res.* **5**, 267–72.
Koshland, D. E., Nemethy, G., and Filmer, D. (1966) *Biochemistry* **5**, 365–85.
Krem, M. M., and Di Cera, E. (1998) *Proteins* **30**, 34–42.
Kristensen, C., Kjeldsen, T., Wiberg, F. C., Schaffer, L., Hach, M., Havelund, S., Bass, J., Steiner, D. F., and Andersen, A. S. (1997) *J. Biol. Chem.* **272**, 12978–83.
Laskowski, R. A., McArthur, M. W., Moss, D. S., and Thorton, J. M. (1993) *J. Appl. Cryst.* **26**, 283–91.
Lau, F. T.-K., and Fersht, A. R. (1987) *Nature* **326**, 811–2.
Lesk, A. M., and Fordham, W. D. (1996) *J. Mol. Biol.* **258**, 501–37.
LiCata, V. J., and Ackers, G. K. (1995) *Biochemistry* **34**, 3133–59.
LiCata, V. J., Speros, P. C., Rovida, E., and Ackers, G. K. (1990) *Biochemistry* **29**, 9771–83.
Lubin, I. M., Healey, J. F., Barrow, R. T., Scandella, D., and Lollar, P. (1997) *J. Biol. Chem.* **272**, 30191–5.
Mann, K. G., Nesheim, M. E., Church, W. R., Haley, P., and Krishnaswamy, S. (1990) *Blood* **76**, 1–16.
Manning, T. C., Schlueter, C. J., Brodnicki, T. C., Parke, E. A., Speir, J. A., Garcia, K. C., Teyton, L. Wilson, I. A., and Kranz, D. M. (1998) *Immunity* **8**, 413–26.
Marki, F., Gasparo, M. D., Eisler, K., Kambler, B., Riniker, B., Rittel, W., and Sieber, P. (1979) *Hoppe–Seyler's Z. Physiol. Chem.* **360**, 1619–32.
Martin, P. D., Malkowski, M. G., DiMaio, J., Konishi, Y., Ni, F., and Edwards, B. F. (1996) *Biochemistry* **35**, 13030–9.
Martin, P. D., Robertson, W., Turk, D., Huber, R., Bode, W., and Edwards, B. F. P. (1992) *J. Biol. Chem.* **267**, 7911–20.
Matthews, B. W. (1993) *Ann. Rev. Biochem.* **62**, 139–60.
Mildvan, A. S., Weber, D. J., and Kuliopulos, A. (1992) *Arch. Biochem. Biophys.* **294**, 327–40.
Mirmira, R. G., and Tager, H. S. (1991) *Biochemistry* **30**, 8222–8229.
Miyata, T., Aruga, R., Umeyama, H., Bezeaud, A. M., Guillin, M. C., and Iwanaga, S. (1992) *Biochemistry* **31**, 7457–62.
Monod, J., Wyman, J., and Changeux, J. P. (1965) *J. Mol. Biol.* **12**, 88–118.
Nakagawa, S. H., and Tager, H. S. (1991) *J. Biol. Chem.* **266**, 11502–9.
Nakagawa, S. H., and Tager, H. S. (1992) *Biochemistry* **31**, 3204–14.

Neurath, H. (1984) *Science* **224**, 350–57.
Ni F., Ripoll D. R., Martin P. D., and Edwards, B. F. P. (1992) *Biochemistry* **31**, 11551–7.
Nicholls, A., and Honig, B. (1991) *J. Comput. Chem.* **12**, 435–40.
Ooi, T., Oobatake, M., Nemethy, G., and Scheraga, H. A. (1987) *Proc. Natl. Acad. Sci. USA* **84**, 3086–90.
Pakianathan, D. R., Kuta, E. G., Artis, D. R., Skelton, N. J., and Hebert, C. A. (1997) *Biochemistry* **36**, 9642–8.
Patthy, L. (1990) *Blood Coagul. Fibrinol.* **1**, 153–66.
Perry, K. M., Onuffer, J. J., Gittelman, M. S., Barmat, L., Matthews, C. R. (1989) *Biochemistry* **28**, 7961–70.
Perutz, M. F. (1989) *Q. Ref. Biophys.* **22**, 139–236.
Pickett, S. D., Sternberg, M. J. E. (1993) *J. Mol. Biol.* **231**, 825–39.
Rawlings, R. D., and Barrett, A. J. (1993) *Biochem. J.* **290**, 205–18.
Rawlings, R. D., and Barrett, A. J. (1994) *Methods Enzymol.* **244**, 19–61.
Rezaie, A. R. (1996) *Biochemistry* **35**, 1918–24.
Richieri, G. V., Low, P. J. Ogata, R. T., and Kleinfeld, A. M. (1997) *J. Biol. Chem.* **272**, 16737–40.
Robinson, C. R., and Sligar, S. G. (1993) *Protein Sci* **2**, 826–32.
Rondard, P., and Bedouelle, H. (1998) *J. Biol. Chem.* **273**, 34753–9.
Sandberg, W. S., and Terwiliger, T. C. (1989) *Science* **245**, 54–7.
Schechter, I., and Berger, A. (1967) *Biochem. Biophys. Res. Commun.* **27**, 157–62.
Scrutton, N. S., Berry, A., and Perham, R. N. (1990) *Nature* **343**, 38–43.
Shirley, B. A., Stanssen, P., Steyaert, J., and Pace, C. N. (1989) *J. Biol. Chem.* **264**, 11621–5.
Shortle, D. (1996) *FASEB J.* **10**, 27–34.
Shortle, D., Meeker, A. L. (1986) *Proteins: Struct., Funct., Genet.* **1**, 81–9.
Shortle, D., Stites, W. E., and Meeker, A. L. (1990) *Biochemistry* **29**, 8033–41.
Smith, G. P., and Petrenko, V. A. (1997) *Chem. Rev.* **97**, 391–410.
Stubbs, M., Oschkinat, H., Mayr, I., Huber, R., Angliker, H., Stone, S. R., and Bode, W. (1992) *Eur. J. Biochem.* **206**, 187–95.
Tidor, B., Karplus, M. (1994) *J. Mol. Biol.* 238, 405–14.
Tsiang, M., Jain, A. K., Dunn, K. E., Rojas, M. E., Leung, L. L. K., and Gibbs, C. S. (1994) *J. Biol. Chem.* **270**, 16854–863.
Vajda, S., Sippl, M., and Novotny, J. (1997) *Curr. Op. Struct. Biol.* **7**, 222–8.
Vila, J., Williams, R. L., Velasquez, M., and Scheraga, H. A. (1991) *Proteins* **10**, 199–218.
Vindigni, A., Dang, Q. D., and Di Cera, E. (1997) *Nature Biotechnol.* **15**, 891–95.
Vindigni, A., and Di Cera, E. (1996) *Biochemistry* **35**, 4417–26.
Warshel, A., Naray-Szabo, G., Sussman, F., and Hwang, J. K. (1989) *Biochemistry* **28**, 3629–37.

Weiner, S. J., Kollman, P. A., Nguyen, D. T., and Case, D. A. (1986) *J. Comp. Chem.* **7**, 230–8.
Wells, C. M., and Di Cera, E. (1992) *Biochemistry* **31**, 11721–30.
Wells, J. A. (1990) *Biochemistry* **29**, 8509–17.
Wesson, L., and Eisenberg, D. (1992) *Protein Sci.* **1**, 227–35.
Young, D. C., Zhang, H., Cheng, Q.-L., Hou, J., and Matthews, D. J. (1997) *Protein Sci.* **6**, 1228–36.
Yu, M.-H., Weissman, J. S., and Kim, P. S. (1995) *J. Mol. Biol.* **249**, 388–97.
Zhang, E., and Tulinsky, A. (1997) *Biophys. Chem.* **63**, 185–200.
Zimm, B. H., and Bragg, J. K. (1959). *J. Chem. Phys.* **31**, 526–35

3

Affinity Prediction: The Sine Qua Non

Garland R. Marshall, Richard D. Head, and Rino Ragno

Significant advances in molecular biology, x-ray crystallography, and NMR spectroscopy provide three-dimensional structures of potential therapeutic targets at atomic resolution at an increasing rate. Coupled with concomitant increases in accessible computing power, and in understanding of the physical chemical basis of molecular interactions, these advances have fostered a new era of rational drug design. Compounds in clinical use as HIV antivirals are a testimony to the effectiveness of iterative structurally based design. Many structurally based drug-design software packages now exist (Bohm, 1992; Ho and Marshall, 1994; Pearlman and Murko, 1993; Rotstein and Murcko, 1993) which assist in the design of novel ligands as potential therapeutics. No matter what algorithmic method of ligand design is utilized, an essential part of the process is the prediction of the affinity for the receptor of the designed ligands, both in their construction and to assist in synthetic prioritization. Affinity for the macromolecular target is, after all, an essential requirement, a sine qua non.

1. Theoretical Considerations

An intellectually satisfying approach to predicting ligand affinity focuses on directly calculating the thermodynamic quantities involved in the formation of a ligand/receptor complex. The fundamental equation underlying affinity relates the free-energy difference (ΔG) between the ligand and the receptor in solution and in the complex to the binding constant (K_a) for the complex: $\Delta G = -kT \ln K_a$, where k is Boltzmann's constant and T is the absolute temperature. At room temperature, this means that a factor-of-ten enhancement in affinity means an approximate 1.3 kcal/mol increase in binding energy. This can be readily decomposed into the changes in enthalpy (ΔH) and entropy (ΔS) of binding: $\Delta G = \Delta H - T\Delta S$. The changes in enthalpy relate to the direct energies of interaction between the atoms of the ligand and those of the receptor (and any differences between internal interactions within the ligand and the receptor in the two states) versus direct interactions with atoms in the solvent. The changes in entropy relate to changes in the ordering of the system. To remove the degrees of freedom relating the position and orientation of the ligand with respect to the receptor (the so-called *cratic* free energy) requires a significant reduction in the entropy of the system (Hermans and Wang, 1997)—on the order of 12–17 kcal/mol often suggested—depending on the molecular weights of the components and the degree to which the relative positions of the atoms are held fixed within the complex. The cratic free energy is a function of temperature, concentration, and molecular weight. For a 1M solution at 25°, a water molecule would lose 8.6 kcal/mol of translational entropy and 3.1 kcal/mol of rotational entropy on binding to a macromolecule, while a drug molecule of molecular weight 200 would lose 10.7 kcal/mol and 8.9 kcal/mol of translational and rotational entropy, respectively. The value often chosen of 14 kcal/mol for the cratic free energy represents an approximate average molecular weight of 60 for the drug molecule. Based on a thorough analysis of the underlying statistical thermodynamics of binding affinities by Gilson et al. (1997), however, it can be argued that molecular mass has negligible effect on the standard free energy of binding. For a detailed review of the statistical thermodynamics of binding, the reader is referred to the paper by Gilson et al. (1997), in which the fundamental theory has been developed sufficiently to resolve a number of disputes in the literature.

The processes govering molecular recognition and affinity consist of transfer of the ligand from the solvent to the generally more hydrophobic protein environment or, in some cases, to the more structured hydrophilic environment. The system undergoes changes in energy and entropy as a result of losing solute–solvent interactions, and the transfer of the ligand from solution to the active site. In most cases, the ligand has to displace water molecules that occupy the active site in its unliganded state. At the binding site, the ligand is conformationally immobilized, thereby decreasing its entropy and increasing its free energy. Simultaneously, optimized interactions with the functionality of the binding pocket increase the enthalpy of binding. The process can involve substantial conformational changes of both the ligand and the receptor which also have an impact of the free energy of binding. An absolute prerequisite for specific binding, however, is steric complementarity. Irrespective of the pattern of intermolecular forces surrounding the ligand, if a bulky group is present that does not fit within the three-dimensional active site available at the receptor, then effective binding is precluded due to the magnitude of the repulsive van der Waals interaction.

In summary, the success of a particular recognition event is dependent upon three major interactions that involve changes in both entropy and enthalpy. The first interaction is the conformational changes of the ligand and the receptor upon complexation. The second is the structural and energetic complementary of the ligand and the receptor in the complex. The third is the thermodynamic aspects that describe the transfer of the ligand from solution to the binding site—namely, the desolvation of ligand and receptor, and the loss of rotational and translational entropy of the ligand. In order to successfully predict the affinities of ligands for receptors, one must attempt to quantify these events and scale their relative contributions to the free energy of binding.

Alternative formulations of the problem can be made, and the free energy dichotomized, in various ways to correspond to experimentally accessible numbers and/or more traditional physical concepts. Williams (Doig and Williams 1992; Searle et al., 1995; Searle and Williams, 1992; Searle et al., 1992; Williams, 1991; Williams et al., 1991) has used an antibiotic (vancomycin)-peptide complex to further evaluate various contributions to binding affinity, and has produced the following simplified relationship which attempts to dichotomize the free energy even further into more-easily recognized components:

$$\Delta G_{binding} = \Delta G_{(trans+rot)} + \Delta G_{rotors} + \Delta H_{conform} + \sum_i \Delta G_i + \Delta G_{vdw} + \Delta G_H.$$

Here, $\Delta G_{(trans+rot)}$ is the cratic free energy associated with translational and rotational freedom of the ligand with respect to the receptor. This has an adverse effect on binding of 50–70 kJ/mol (12–17 kcal/mol) at room temperature for ligands of 100–300 daltons assuming complete loss of translational and rotational freedom relative to the receptor. One of the sources of error in many calculations is the estimation of the percent loss of these degrees of freedom, as this can vary greatly between systems under study. The ΔG_{rotors} term is the free energy associated with the number of rotational degrees of freedom frozen. This is 5–6 kJ/mol (1.2–1.6 kcal/mol) per rotable bond, assuming complete loss of rotational freedom—again a value that is unlikely and system-dependent. The $\Delta H_{conform}$ term is the strain energy introduced by complex formation (deformation in bond lengths, bond angles, torsional angles, and so on, from the solution states of both ligand and receptor). The sum $\sum \Delta G_i$ is the total interaction free energies between polar groups. The ΔG_{vdW} term is the energy derived from enhanced van der Waals interactions in the complex. Finally, ΔG_H is the free energy attributed to the hydrophobic effect (12.5 J/mol per square nanometer of hydrocarbon surface removed from solvent by complex formation, related to the free-energy changes, both enthalpic and entropic, when water is released from a hydrophobic interface and analogous to a potential of mean force).

As several of the components in the binding-energy estimate are directly related to the degree of order of the system (entropy), simulations in solvent may be necessary to quantify the degree by which the relative motions of the ligand and protein are quenched, and to determine the restriction on rotational degrees of freedom upon complexation. Hermans and Wang (1997) have shown how one can determine these values accurately, using molecular dynamics, through the use of novel potentials that constrain the rotational and translational degrees of freedom of the ligand. One should, therefore, consider such thermodynamically rigorous approaches in which a ligand of known structure and affinity is mutated to the ligand of interest and the difference in binding free energy is calculated (Hermans and Wang, 1997; Jorgensen and Ravimohan, 1985; Kollman, 1993; Straatsma and McCammon, 1991). A good example of the use of such simulations to calculate the relative free energy of binding is that of the binding of benzene to the enzyme T4 lysozyme (Hermans and Wang, 1997). This approach has recently been critically reviewed by Gilson et al. 1997. Such methods are powerful when limited structural variation is under consideration, since they have been shown accurate when the perturbation to the starting ligands is small. Unfortunately, such thermodynamic-cycle perturbation approaches are computationally demanding and currently inappropriate for de novo design, due to the number and diversity of structures to be considered. The basic concepts underlying such an approach is illustrated below in the discussion of free-energy perturbation methods. Åqvist has used limited molecular dynamics simulations to estimate the entropic changes on complexation in a scoring function shown to be useful for several complexes of therapeutic interest. (Åqvist et al., 1994; Hansson et al., 1998; Hulten et al., 1997). Paulsen and Ornstein 1997) have examined this approach with a series of 11 substrates of cytochrome P450cam. Gorse and Gready (Cummins and Gready, 1993a,b; Gorse and Gready 1997) have taken a similar approach to calculating the affinities of a congeneric series of inhibitors of dihydrofolate reductase. Marelius et al. (1998) have applied the Åqvist method to DHFR for a different series of inhibitors.

2. Free-energy Perturbation Methods

Because of the difficulties in sampling configurational space to generate an accurate estimate of the partition function for the free ligand and receptor, as well as for the complex, the preferred methods are those that focus on the difference $\Delta\Delta G$ in free energy of binding between a known ligand with an experimental measurement of affinity and another ligand of similar structure. Using a closed thermodynamic cycle, molecular dynamics, and the experimental binding free energy of one inhibitor I)I_1 (ΔG_1) the method of free-energy perturbation (FEP; also known as

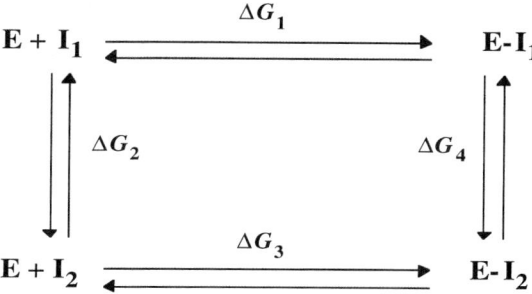

Figure 3.1. Thermodynamic Cycle

thermodynamic-cycle perturbation) permits one to calculate the binding free energy of new inhibitors I_2.

Considering the cycle in fig. 3.1., it is possible to write the relation

$$\Delta G_1 - \Delta G_3 = \Delta G_2 - \Delta G_4,$$

from which ΔG_3 can be obtained:

$$\Delta G_3 = \Delta G_2 + \Delta G_4 - \Delta G_1.$$

The values of ΔG_1, ΔG_2, and ΔG_4 can be calculated using molecular dynamics. The simulation of ΔG_1 and ΔG_2 is laborious, since there is a need to model not only the binding event but also the desolvation processes of both the ligand (I) and the receptor (E).

In calculating ΔG_2, the inhibitor I_1 is gradually transformed into I_2 (perturbation) during the course of the simulation in the solvent. The quantity ΔG_4 is also calculated using a very similar method, but the molecule of the ligand is held within the active site (bound state) where the effect of the receptor is now taken into account. The overall process of the perturbation is performed in a number of small steps, starting from the inhibitor (ligand) I_1 described with its own parameters at 100% and then changing slowly the parameters into those belonging to the other inhibitor I_2. In order to allow the system to accommodate the changes in inhibitor structure at each step, the whole perturbation must be fairly small, namely by substituting only a single atom (changing hydrogen to fluorine, for example), for this procedure to be successful. Furthermore, a long period of simulation is required to ensure the full equilibration of the changing system—for example, 100 ps. Finally, the simulation should also be performed in the reverse direction to check for consistency.

There are several examples of successful application of this method in studies of HIV-1 protease inhibitors. Ferguson et al. (1991) used the FEP method (Amber force field and TIP3P water) to ascertain whether the binding affinity of the S or the R form of the transition-state hydroxyl of the HIV-1 protease inhibitor JG365 was the most favorable to complex formation. Their results, when compared with parallel and independent work on experimental determinations, showed good agreement, indicating the S diastereomer to be the most active ($\Delta\Delta G_{exp} = 2.6$ kcal/mol, $\Delta\Delta G_{calcd} = 4.0$–$2.8$ kcal/mol). Furthermore, the same simulations suggested that the HIV-1 protease is singly protonated on the active-site carboxyls between Asp 25 and Asp 125. In a very similar manner, but using a different program, force field (CEDAR), and water model (SPC) (Tropsha and Hermans, 1992), Chen and Tropsha (1995) carried out a simulation on the same molecular system (JG354/HIV-1_Pr) obtaining similar results ($\Delta\Delta G_{calcd} = 2.7$ kcal/mol). Rao and Murcko et al., 1992) also used the FEP approach in order to rationalize the novel binding mode of a potent HIV-1 protease inhibitor, Ro 31-8959, whose R configuration of the central transition state hydroxyl showed binding ($IC_{50} < 0.4$ nM) about 250-fold better than the S diastereomer ($IC_{50} > 100.0$ nM). The S diastereomeric hydroxyl is normally observed to bind tighter as the case of JG365, where the S configuration is preferred by 2.6 kcal/mol. The differences in the preferred stereochemistry of the hydroxyl had previously been rationalized by Rich et al. (1991) based on molecular modeling studies that were later confirmed when the crystal structures of the two complexes became available.

Reddy et al. (1991) combined molecular mechanics, dynamics, thermodynamic-cycle perturbation calculations, molecular design, synthesis, biochemical testing, and crystallographic structure determination of complexes in an elegant way, to construct an iterative computer-assisted drug-design approach to predict the binding affinity of novel compounds of HIV-1 protease. By this approach, they were able to design new structures by significant mutation of the C-terminal Val-Val mthyl ester of a known hydroxyethylene inhibitor (Graves et al., 1991) to a diphenhydramine amide derivative in which two phenyl moieties fill the p2' and p3' side-chain pockets in the HIV-1 protease. Subsequent calculation led to the replacement of a phenyl ring with an indole. Synthesis and biochemical testing confirmed the theoretical results showing a better inhibition potency for the indolyl derivative (0.2 mM vs 1.7 mM of the diphenhydramine derivative).

Helms and Wade (1998) have reported recently a new variant of the free-energy perturbation approach to calculating binding affinity. In their study of the binding of camphor to P450cam, they mutated the ligand to six water molecules in the binding site. The computed ΔG of binding was within 3 kJ/mol of the experimental value. This was a well-selected case, since the binding pocket is buried and isolated from bulk solvent.

Due to the computational demands of a thorough investigation of binding affinity, several groups have explored simplified approaches that utilize simulations to help approximate the thermodynamics of the system under study. The groups of Åqvist (Åqvist and Warshel, 1989; Hulten et al., 1997) and of Gready (Cummins and Gready, 1993a,b; Gorse and

Gready, 1997) have used simulations to estimate the changes in entropy of the system. Hulten et al. (1997) used the approach of Åqvist to estimate the binding affinities of 10 cyclic HIV-1 protease inhibitors that were systematically overestimated by approximately 2 kcal/mol. Radmer and Kollman (1998) have explored three approximate methods to estimate the affinity of ligands that were not simulated. Each approach reproduced the general trends in binding free energies, suggesting that they might be useful in drug design, but again perturbations to the known complex structure of the ligand were minimal.

3. Heuristics in Calculation of Affinities

Quantitative structure–activity relationships (QSAR) represent another approach to affinity prediction. In these methods, the thermodynamics of binding are not explicitly represented, but are embedded in physicochemical properties determined for each ligand, which are then correlated to activity. In other words, a training set of ligands whose binding affinities are known are used to train a model based on correlation with a set of physicochemical properties, either measured or calculated for each ligand. Traditional QSAR methods have been developed for receptors with unknown three-dimensional (3D) to analyze a database of ligands whose structures and activities (binding affinities against the purified target in the ideal case) are known (Martin, 1978). Such analyses are generally based on the assumption that a correlation exists between the enthalpy of binding and the free-energy of binding, since the receptor is a constant and the congeneric series of ligands under consideration do not differ significantly in size, flexibility, and other features that would impact the entropy of binding. The difficulty with such methods lies in the limitations on the predictive models they generate and the requirements for their appropriate use. First, a considerable training set of ligands of diverse structure with known binding affinities must be present for each receptor. New therapeutic targets that are of the most interest in the pharmaceutical industry generally lack such a set of diverse ligands with measured affinities. Second, the accuracy of prediction of ligands has been generally shown to depend on the similarity of the sought ligand to those in the training set (Folkers et al., 1993); that is, interpolation generally gives superior results to extrapolation. Thus, there is little confidence in the prediction of binding affinity of a novel ligand that is truly unique with respect to the training set, due to the extrapolation required from ligands present in the training set.

Three-dimensional QSAR methods, such as comparative molecular field analysis (CoMFA—Cramer et al., 1988; Green and Marshall, 1995; Marshall and Cramer, 1988) or the hypothetical active-site lattice (HASL) approach (Doweyko, 1988), often use a grid-based approach to derive an active-site model for the receptor in terms of energetic fields such as electrostatics, sterics, and hydrophobicity (Abraham and Kellogg, 1993). In an extension to receptors of known structure, 3D QSAR has been used to derive robust predictive models for the binding affinity of inhibitors of HIV protease (Oprea et al., 1994 a,b; Waller and Marshall 1993; Waller et al., 1993), but one can argue that most of these compounds are of similar nature, being derived by substrate modification and retaining essentially a peptidic backbone.

Previous approaches have utilized a training set specific to each therapeutic target, in order to derive a predictive 3D QSAR model. In many cases, an appropriate set of compounds of known affinity and of sufficiently diverse structures became available in time only for a retrospective analysis with therapeutic candidates alrady in clinical trials. The efficient use of de novo design for a 3D structure of a novel therapeutic target requires the ability to predict affinity without waiting for a training set. Sampling problems associated with molecular-mechanical simulations preclude their use, except for very small perturbations in structure, which renders this approach virtually useless for de novo design.

In an effort to overcome the limitations and approximation of QSAR approaches, a new class of scoring function with greater range of applicability has been developed. This type of function uses the receptor's 3D structure along with that of the ligand to predict affinity (Bohm, 1994; Holloway et al., 1995; Horton and Lewis, 1992). Thus, an empirical scoring function which can approximate the affinity of a ligand by examining the probable complex with the target is clearly desirable, and several groups (Bohm 1992; Eldridge et al., 1997; Miranker and Karplus, 1991; Nishibata and Itai, 1991; Rotstein and Murcko 1993; Weng et al., 1997) have attempted to develop such a reliable function.

Such approaches—for example, the scoring function for LUDI (Bohm, 1994)—base their calculations on an estimate of the binding free energy by approximating the contributions of hydrogen bonding, of entropy due to rotatable bonds in the ligand that are frozen upon binding, and of desolvation based on some sort of hydrophobic complementarity information. Bohm (1994) analyzed 45 protein–ligand complexes (affinity range from 9 to 76 kj/mol) and found the following equation by multiple regression analysis:

$$\Delta G_{binding}(kj/mol) = 5.4 \Delta G_0 - 4.7 \Delta G_{hb} - 8.3 \Delta G_{ionic} - 0.17 \Delta G_{lipo} + 1.4 \Delta G_{rot},$$
$$r^2 = 0.76, \quad S = 7.9, \quad q^2 = 0.696, \quad s_{press} = 9.3 \text{ kJ/mol (2.2 kcal/mol)},$$

where ΔG_o is related to the reduction in rotational and translational entropy, ΔG_{hb} is the free energy associated with hydrogen-bond formation, ΔG_{ionic} is the binding energy from ionic interacations, ΔG_{lipo} is the lipophilic interaction contribution, and ΔG_{rot} is the energy loss by freezing of internal degrees of freedom in the ligand.

Similarly, Krystek et al. (1991) analyzed 19 protein–ligand complexes in an update of the Novotny approach to binding entropy (Novotny et al., 1989) and produced the following relationship:

$$\Delta G_{\text{binding}}(\text{kcal/mol}) = 11 - 0.025\Delta G_{\text{CSA}} - \Delta G_{\text{El}} + 0.6 T_{\text{sc}},$$
$$r^2 = 0.69, \quad S = 4.0,$$

where ΔG_{CSA} is the hydrophobic energy due to loss of surface area, ΔG_{El} is the electrostatic binding energy, and T_{sc} is the change in torsional and rotational entropy.

Eldride et al. (1997) have reported a scoring function to be used in conjunction with de novo design software to prioritize candidates. The scoring function consisted of the hydrogen-bond function from LUDI, an atomic lipophilic function, a term for frozen rotatable bonds, and a term for metal interactions. The coefficients for these terms were obtained by multiple linear regression based on a training set of 82 complexes whose crystal structures were known. The cross-validated q2 of the training set was 0.658, and the standard error of prediction was approximately 2.1 kcal/mol.

DeWitte and Shakhnovich (1996) have developed a knowledge-based potential in connection with their de novo design method SMoG to give an estimate of the binding affinity. Three systems were examined for which experimental data was readily available. Binding data from two systems—HIV protease and purine nucleoside phosphorylase—could be correlated ($r^2 = 0.77$) with estimated free energy, while data from the SH3 domain required a different slope and intercept. This implies that the simple interaction potential does not fully represent the relevant interactions contributing to affinity.

Vajda et al. (Vajda et al., 1994; Weng et al., 1997) have calibrated an empirical free-energy function by assuming that binding does not effect the conformational free energy of either ligand or receptor. By further assuming that the vdW interactions between solvent and free ligand and receptor basically balanced those found in the complex, the function could be reduced to a calculation of electrostatics and desolvation. The average difference between the calculated and measured affinities for a series of protease inhibitors was approximately 1.3 kcal/mol (Vajda et al., 1994), implying that the entropic changes in the ligands were similar upon binding.

Even these methods of approximating affinities have their limitations. First, they require the 3D structure of the receptor to make their predictions, although this is becoming less of a problem for many systems of therapeutic interest. Second, most of these methods employ calculations of limited accuracy due to computational limitations on the estimation of the entropic contributions. New hybrid methods focus on the accuracy of the calculations through maximal use of 3D information of the ligand–

receptor complex by combining a heuristic approach, to deal with entropic factors, with parameters derived from molecular mechanics that focus on enthalpic issues. Based on their experience with 3D QSAR, Head et al. (1986) realized that the necessary information is inherent in the complex, and that an appropriate correlation between calculable physical chemical parameters and the affinity of the ligand should be possible. In other words, the enthalpy of interaction can be reasonably estimated by molecular mechanics, but quantitative entropic estimation is much more difficult, and heuristics might be able to provide adequate accuracy. Head et al. chose to study the crystal structures of complexes of a variety of structural types (enzyme–inhibitor, protein–protein, antibody–antigen, etc.) as a training set for a generalized correlation function, VALIDATE, to predict the affinity of novel complexes of ligands with receptors without a specific training set for the receptor.

Parameter sets that were chosen to examine included a variety of terms related to binding, including a number that reflect desolvation of the ligand and receptor (entropic). Those parameter sets included both steric and electrostatic interaction energies, $H \log P$, a steric fit term to count the direct contacts between ligand and receptor, terms for the complementarity of lipophilic and hydrophilic contact between ligand and receptor, internal degrees of freedom in the ligand, and a conformational enthalpy (strain) term. The choice of parameters is further discussed below. Either PLS or neural networks were used as the training paradigm, and the cross-validated q^2 used to evaluate the models derived. A conceptually similar approach by Ortiz et al. (1995) used parameters derived from molecular-mechanical calculations on the ligand–receptor complex, and linear regression analysis, to develop a model of the affinity of ligands for a single enzyme (a phospholipase).

4. Changes in Internal Entropy

Searle and Williams (Searle and Williams, 1992; Searle et al., 1992) have examined the thermodynamics of sublimation of organic compounds without internal rotors, and shown that only 40–70% of theoretical cratic entropy loss occurs on crystallization. This provides an estimate of the entropy loss (20–50 kJ/mol) to be expected on small drug–ligand interaction and a potential error of prediction of over five orders of magnitude in affinity. Thus, one critical aspect in the calculation of affinity is the means by which entropic changes are calculated or estimated.

Changes in conformational entropy occur when the freely rotating side chains of the dissociated components are forced to adopt more rigid conformations on complex formation. Pickett and Sternberg (1993) have analyzed the side-chain conformational entropy change on protein folding for each type of amino-acid residue. Novotny et al. (1989) attempted to deal with entropy changes in free-energy calculations of antibody–antigen complexes. The minimal estimate of conformational freedom lost assumed that each torsional degree of freedom has approximately three equivalent energetic states available, namely, the trans and ± gauche. To estimate the total change in side chain conformational entropy (ΔS_{CF}), the atoms involved in the contact area of the complex were used to estimate the number, N, of sidechain torsions fixed:

$$\Delta S_{CF} = -R \ln 3^N = -NR \ln 3 = -2.18 N \, \text{kJ/mol}.$$

Williams et al. (1991) have estimated the entropy change due to freezing a free rotor to be 5 kJ/mol. Similarly, an indication of the conformational entropy lost upon binding (Murphy et al., 1994) can be calculated by estimating the flexibility of the ligand. The flexibility index developed by Fisanick et al. (1990) is a function of the shortest topological paths between all pairs of atoms in a structure, and takes into account the types of bond and the extent of branching in the paths. An alternative measure of this is simply to count rotatable bonds that would have reduced rotational freedom in the complex. Obviously, the degree to which the free rotation is impeded depends on the location of the rotor within the complex, and some rotors will be more impeded than others. Estimates of these differences can be obtained by limiting simulations that form the basis of the approach of Åqvist et al. (1994) to calculate affinities.

In VALIDATE (Head et al., 1996), the number of rotatable bonds in the ligand which are frozen on interaction with the receptor are counted by summing all nonterminal single bonds (methyl groups are assumed to be freely rotating). It has been suggested that the number of degrees of freedom in a nonaromatic ring system (aromatic rings are ignored as having no internal rotational degrees of freedom) is of the order $n - 6$ (Go and Scheraga, 1970), where n represents the number of bonds in the ring. In VALIDATE, $n - 4$ was chosen as more appropriate, based upon the conformational analysis of five-membered systems such as proline. To count such a ring as having zero degrees of freedom did not seem appropriate, since it could pucker.

Therefore, the number of rotatable bonds was expressed as

$$\text{rb} = 1 * (\text{number of nonterminal single bonds}) + \sum_i (n_i - 4),$$

n_i = number of bonds in ring i,

where the count is restricted as follows. In protein–protein and even small ligand–protein systems, some portion of the ligand is not bound at the active site and remains freely accessible to the solvent. Rotatable bonds in these areas, therefore, cannot be considered as being more "frozen" by complex formation. Hence, only the rotatable bonds at the active-site interface are counted.

5. Overall Partitioning from Solvent to Binding Site

The energetic gains arising from desolvation of both ligand and receptor upon complexation can be approached from a variety of viewpoints, as exemplified in the different scoring functions. Jackson and Sternberg (1995) have developed a scoring function for docking proteins which uses changes in molecular surface area, in combination with electrostatic free energy and side-chain conformational entropy changes, to distinguish near-native from nonnative dockings of six protein complexes. To illustrate an approach, the rationale underlying VALIDATE is presented. From a physicochemical viewpoint, the lipophilicity of a hydrophobic molecule is estimated as the energy needed to create a cavity in the aqueous solvent in which the solute can fit. With hydrophobic molecules, when the molecule binds to the receptor, the energy of cavity creation for the solvent surrounding the ligand is released, entropically favoring the recognition process. The (base-10) logarithm of the partition coefficient P is used as a measure of molecular lipophilicity. VALIDATE (Head et al., 1996) uses the fragment-based $H \log P$ (Hansch and Leo, 1979) method in Hint 1.1 (Kellogg et al., 1991) to compute the ligand's partition coefficient. With the partition coefficient, a negative value indicates a preference for a polar (hydrophilic) environment, and a positive value indicates a preference for a non polar (lipophilic) environment. It is not sufficient to use simply the partition coefficient without using the additional knowledge of the active site of the receptor. For example, the HIV protease active site, though charged with several polar groups, is predominantly lipophilic in terms of total surface area. Thus, ligands with a positive partition coefficient are favored for the site binding HIV protease. However, in the case of the L-arabinose sugar binding protein, most of the active site is hydrophilic, as a proportion of its total surface area. The sugar ligands that bind to this protein are hydrophilic (negative partition coefficients). In VALIDATE Head et al., 1996), the amount of hydrophilic and lipophilic surface area was computed as ratios to the total surface area of the receptor active site. The use of the partition coefficient (PC) was then modified according to this information in the following fashion:

$$PC = RC * H \log P_p C,$$

where HlogP_PC is the computed partition coefficient, and

$$RC = \begin{cases} 1 \text{ if the receptor's active site is predominantly lipophilic,} \\ -1 \text{ if the receptor's active site is predominantly hydrophilic.} \end{cases}$$

Determination of the lipophilic/hydrophilic preference of the receptor active site in VALIDATE was more involved than simply comparing surface areas. The calculation of lipophilic and hydrophilic surface area in the active site was done in VALIDATE (Head et al., 1996) using each ligand that binds with the particular receptor as a seed to focus on the relevant interacting surface. The ligand was placed in the binding site, and each receptor atom which has a solvent-accessible surface that was within a distance R (where R is the mean radius of the solvent, water with a mean radius of 0.14 nm (nanometers) was used for the calculations in this paper) of the solvent-accessible surface area of any ligand atom was considered part of the active site. The solvent-accessible surface area of this atom was added to the appropriate surface-area type. Types were defined as follows: any carbon that is covalently bonded to no more than one non carbon is considered lipophilic; any hydrogen connected to such a carbon is also considered lipophilic; all other atoms are considered hydrophilic. This is essentially the definition used in the Bohm scoring function (Bohm, 1994). Only a small portion of a protein ligand, the active site, is desolvated by binding to the receptor; only the $H \log P$ for this region is relevant. Therefore, the calculation was done only on this part of the molecule.

6. Steric and Electronic Complementarity

An absolute prerequisite for specific binding is structural complementarity of ligand and receptor at the site of interaction due to the steep repulsive nature of the van der Waals interactions when atoms are too close. Due to the competitive interactions with water in the unliganded state, there must be interaction between complementary desolvated surfaces, such that polar atoms are properly positioned to make hydrogen bonds. That is, electrostatic interactions contribute to the specificity of the complex formation; incorrect associations are forbidden by large unfavorable enthalpies, due to poor packing and loss of hydrogen bonds switched to water. The nonbonded electrostatic interaction energy is calculated using the explicit sum of the coulombic potentials:

$$\Phi = \frac{1}{4\pi\varepsilon_0} \sum_i^L \sum_j^R \frac{q_i q_j}{r_{ij}}.$$

In VALIDATE, the charges on the ligand and the receptor were those from the implementation of the Amber force field within the MacroModel (Mohamadi et al., 1990) program. Additions to this Amber charge set in MacroModel allowed for appropriate charges for non-amino-acid structures. The nonbonded steric interaction energy was computed from the explicit sum of the Lennard–Jones potentials.

$$E_{\text{vdw}} = \sum_i^L \sum_j^R \varepsilon_{ij} \left(\frac{1}{R_{12}} - \frac{2}{R_6} \right),$$

where

$$\varepsilon_{ij} = \sqrt{\varepsilon_i \varepsilon_j},$$

$$R_{12} = \left(\frac{r_{ij}}{R_i + R_j}\right)^{12}, \quad R_6 \left(\frac{r_{ij}}{R_i + R_j}\right)^6;$$

here, r_{ij} is the distance between atom center i and atom center j, and R_i and ε_i (R_j and ε_j) are the respective van der Waals radius and epsilon value of atom i (atom j).

All parameters required were derived from the Amber force field within the MacroModel program.

7. Complementary Surface Areas

In order to estimate more accurately the desolvation energies and enthalpies of ligand–receptor interaction, estimates of the interaction surfaces in the complex and their nature are often made. In this regard, VALIDATE computed four components to surface complementarity. These were lipophilic complementarity (nonpolar–nonpolar), hydrophilic complementarity (polar–polar, opposite charge), lipophilic–hydrophilic (polar–nonpolar) noncomplementarity, and hydrophilic (polar–polar, like charge) noncomplementarity. VALIDATE used 256 evenly distributed data points, obtained from the SASA program (Le Grand and Merz, 1993), which were placed on the van der Waals surface of each receptor atom whose van der Waals surface was within 0.5 nm of the atom center of any ligand atom. If a point on this surface was within a mean solvent radius (0.14 nm for water) of the van der Waals surface of a ligand atom, it was considered a contact point. Two different types of calculation of surface area were considered. The first was an absolute surface area between ligand and receptor similar to the method used by Bohm (1994). For each point on the receptor surface, a record of each type of contact was kept. However, a particular type of contact, e.g. lipophilic complementarity, was counted only once, even if that point was within the distance limit described above of more than 1 ligand atom's van der Waals surface. The points of each type were then summed for each atom in the receptor. The total surface on each atom for each type of contact was computed by dividing the number of contact points of that type by 256 (the total number of points possible) and then multiplying the total surface area of the atom. The functional form is

$$\mathrm{CSA} = \sum_i^R \frac{4\pi r_i^2 * \mathrm{CP}_i}{256}$$

where CP is the number of contact points on atom i, and r_i is the van der Waals radius of atom i.

The second method was somewhat similar to the approach in Hint 1.1 (Kellogg et al., 1991)—a pairwise sum. If a single point on the surface of a receptor atom was within the described distance of n atoms of the ligand of a given type, then that point would be assigned a sum of n, rather than 1 as previously described. While one might expect one or the other procedure to be preferred, using both sets of data in the model improved the results with VALIDATE. Clearly, this is considerable cross-correlation in these measurements, and VALIDATE II (below) has reduced these parameters.

8. Estimates of Ligand Strain Energy

Ligand-strain energy may be viewed as the amount of entropy required for the ligand to adopt the receptor-bound conformation. One can defined the strain energy pragmatically by

$$IE = |E_{bs} - E_{solv}|,$$

where E_{bs} is the energy of the ligand's receptor-bound conformation, and E_{solv} is the energy of the ligand in solvent at its nearest local minimum.

In VALIDATE, this was estimated by comparison of the conformational enthalpy of the receptor-bound conformation of the ligand to the nearest local minimum of the unbound ligand using the GB/SA (Still et al., 1990) solvation model with the Amber all-atom force field as implementation in MacroModel.

9. Heuristic Analyses and Results

The parameters described previously and computed by VALIDATE were placed into a spreadsheet contained the binding affinity which was used as the dependent column of a PLS analysis (Wold S et al., 1984 1993). The analysis produced a coefficient for each parameter that was used for the prediction of the binding affinity of new compounds. For comparison, an analysis with the SONNIC neural-network program (Green and Marshall, 1995) with the same parameters as input was also done. The results of both methods were essentially identical, implying that no significant nonlinear correlation in the training parameter data was detected.

The training set used in the VALIDATE paper consisted of the parameters calculated for 51 receptor–ligand co-crystalline complexes. Most of the complexes were selected from a list compiled by Keske and Dixon (1995), which contains co-crystal complexes, available in the Brookhaven database, along with their published binding affinities. The ligands in this training set ranged from 24 to 1512 atoms, and the activity range extended from $-\log K_i = 2.47$ to 14.0. The statistical (PLS) and neural-network (SONNIC) analyses produced similar results with respect to the correlation values. The PLS analysis yielded $r^2 = 0.849$ with $s = 1.006$ log units. A leave-one-out crossvalidation of the set produced $q^2 = 0.776$ with $s_{press} = 1.139$ with six components: six components had been chosen based on previous experience with these parameters as being the most robust. For this particular training set, s_{press} was equivalent for the four-component model. The SONNIC analysis trained to $r^2 = 0.81$. Crossvalidation yielded $q^2 = 0.765$ at this level of training and root-mean-square 1.08. The quantitites r^2, q^2, s, and s_{press} are defined as follows:

$$q^2 = 1 - \left(\frac{\sum_{i=1}^{N}(\text{pred}_i - \text{actual}_i)^2}{\sum_{i=1}^{N}(\text{actual}_i - \overline{X})^2} \right), \quad s - \sqrt{\frac{\left|\sum_{i=1}^{N}(\text{calc}_i - \text{actual}_i)\right|}{N - c - 1}},$$

$$s_{press} = \sqrt{\frac{\left|\sum_{i=1}^{N}(\text{pred}_i - \text{actual}_i)\right|}{N - c - 1}},$$

where N is the number of molecules and c is the number of PLS components (in the case of multiple regression analyses, c is the number of variables).

The true measure of any model rests in its ability to predict. Separate test sets were compiled from recent literature to test the derived model's predictiveness derived with VALIDATE. The first test set consisted of 14 inhibitors that were obtained from crystalline receptor–ligand complexes (fig. 3.2). Neither ligands nor the specific receptors in this test set were included in the training set. Prediction of the binding affinities for this test set of novel complexes (the mean binding affinity is 6.90, against 7.495 for the training set) was very good, with a predictive r^2 of 0.806, an absolute average error of 0.697 (approximately 0.95 kcal/mol at 25° C), and a root-mean-square error of 0.899 log units (approximately 1.2 kcal/mol at 25°C). On this test set, SONNIC scored a maximal predictive r^2 of 0.85 and an absolute average error of 0.96 log units for training to a value of $r^2 = 0.84$ for the training set. A second test set consisted of 13 HIV protease inhibitors whose initial conformation and alignment were derived from the CoMFA analysis done by Waller et al. (1993). The PLS predictive r^2 was 0.568 with an absolute average error of 0.726 log units and a root-mean-square error of 0.866 log units (approximately 1.17 kcal/mol at 25°C). The SONNIC analysis yielded similar results, with a maximal predictive r^2 of 0.529 and an absolute average error of 0.718 log units at a training r^2 of 0.83. The predictive r^2 is considerably lower that that of the first test set; however, this is due to the smaller range and similar values to the mean activity of the training set. The absolute average error is almost

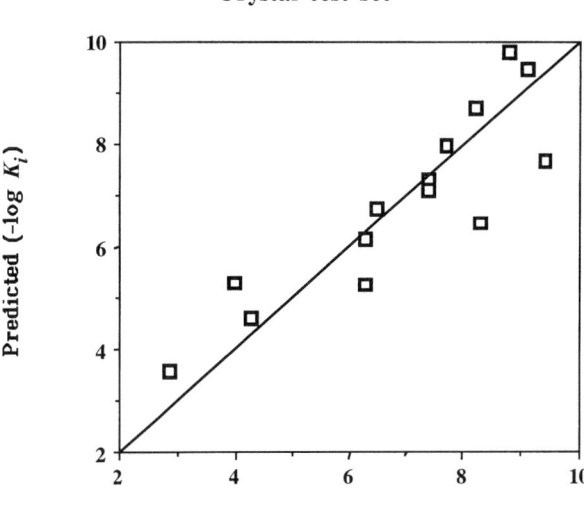

Crystal test set

Figure 3.2. Prediction of affinities of 14 crystalline complexes using coefficients from fitted PLS analysis of the training set. The predictive $r^2 = 0.806$ and the absolute average error = 0.697.

identical to that of the first test set. This example illustrates a new approach for the prediction of binding affinity, based on a hybrid model combining enegetic considerations from molecular mechanics and calculated molecular properties related to desolvation and entropy loss on binding. The similarity between the results of the PLS and neural-network analysis is a significant indication that the derived model is consistent and reflective of information contained within the data, since the results are independent of methodology.

10. Further Heuristic Efforts (VALIDATE II)

The results obtained with VALIDATE were sufficiently encouraging for Marshall and Ragno (manuscript in preparation) to have further explored the energetic description of receptor–ligand interactions. They studied inhibitor complexes with HIV protease to reduce the systematic errors associated with compounds assayed by a variety of methods and concerns over alternate binding modes. Twelve new physicochemical parameters were added to those of VALIDATE to generate a second version, VALIDATE II. It was hoped that choice of an appropriate parameter set might contain more concise representations—that is, increase the signal-to-noise ratio—of the relevant physicochemical phenomena involved with binding. A training set of 49 HIV protease–inhibitor crystalling complexes were used to derive a predictive model, of which the predictiveness was tested on 31 test sets of modeled HIV protease–inhibitor complexes derived from the literature, for a final global test set of almost 400 modeled complexes. The inhibitors in the test set were simply compared with the compounds in the test set, and the closest crystalline analog was used as a template for sketch and alignment followed by minimization with MacroModel.

Variable selection by GOLPE (Baroni et al., 1993) performed on the experimental training set led to a final model with only nine explicative variable and two principal components. Variable selection on a matrix with only the original VALIDATE variables did not lead to an improved model, as it did with the increased number of variables used in VALIDATE II.

A possible nonlinear dependence of the biological activity with the lipophilicity expressed by the log partition coefficient (log P)) has been widely shown (Kubinyi, 1993). Thus in VALIDATE II, the squared Hint-calculated partition coefficient ($H \log P^2$) was also considered. While the $H \log P$ introduced in VALIDATE is an estimate of the lipophilic/hydrophilic preference of the ligand, in VALIDATE II the free energy of solvent cavity formation is taken as a direct measure of the amount of energy released when the ligand goes from the free-solution state to the receptor-bound state. In VALIDATE II, the energetics of the ligand desolvation are estimated by the free energy of solvent-cavity formation obtained from a 1SCF calculation performed on the ligand structure using the AM1 Hamiltonian with solvation model 2 (SM2) implemented in AMSOL 4.0. The inhibitor conformation adopted was that obtained from the GB/SA minimization performed with MacroModel to compute the ligand strain energy.

11. Overall Electrostatics of the Ligand

In VALIDATE II, the HOMO (Highest Occupied Molecular Orbital) energy is used to estimate the potential electronic interaction of the ligand. This parameter is related to the ionization potential of a given structure, and thus measures the tendency of the molecular to interact with either a water molecule or a fragment of the protein receptor pocket.

12. Overall Size and Shape of the Ligand

The size of a molecular is an important historical parameter used in classical QSAR (i.e., Hansch analysis), for example, using STERIMOL parameters or molar refractivity (MR). In VALIDATE II, the molecular weight and the molecular volume are used to estimate the molecule's bulkiness.

These two parameters are readily computed in the same AMSOL calculation of the free energy of solvent-cavity formation. Due to the fact that GOLPE (Baroni, et al.,1993) was used to extract the more informative parameters leading to the best relationship, the total lipophilic and hydrophilic surface areas and the overall contact surface areas were also included as parameters in VALIDATE II.

12. Estimates of Receptor Strain Energy

The receptor strain energy (RSE) can be imagined as the amount of energy required for the receptor to adopt the conformation seen in the complex. The receptor strain energy was pragmatically estimated by

$$RSE = E_{bs} - E_{min},$$

where E_{bs} is the energy of the receptor in the bound state, and E_{min} is the energy of the receptor at its nearest local minimum. In VALIDATE II, this is calculated by comparison of the conformational enthalpy of the bound-state conformation of the receptor to the nearest local minimum of the unbound state using the Amber all-atom force field implementation in MacroModel (Mohamadi et al., 1990) in vacuum. Considering that the HIV-PR binding site is very flexible (flaps opened in the unbound state and closed in the bound state (Collins et al., 1995)), a parameter to estimate such behaviour could greatly improve prediction of the binding affinity.

13. Hydrogen Bonds Between Ligand and Receptor

The interaction of two or more electronegative atoms linked by a hydrogen atom is called a hydrogen bond. In biological systems, the most important hydrogen bonds are those involving the oxygen and nitrogen atoms of the carboxyl, hydroxyl, carbonyl, amino, and amido groups, which are responsible for maintaining the tertiary structure of protein and nucleic acids, as well as the binding of many drugs. Other atoms occasionally involved in drug-receptor hydrogen bonding are sulfur, phosphorus, fluroine, and other halogens, all of which can function as either donors or acceptors of protons. Empirical estimates of hydrogen-bond strength in water are of 1–3 kcal/mol, while the corresponding value derived from the observed heats of sublimation of amide crystals is 5 kcal/mol (Colman et al., 1987). In VALIDATE II, the number of hydrogen bonds formed between ligand and receptor is simply counted. To determine this number, the program HB-PLUS (McDonald and Thornton 1994) was modified to calculate only the number of hydrogen-bond interactions occurring between ligand and receptor. Since HB-PLUS reads PDB file format, a routine to convert the SYBYL MOL2 files of the complexes generated in VALIDATE II was written.

14. Hydrogen Bonds Between Ligand and Solvent

The capacity of ideal hydrogen bonds between the molecule and the environmental solvent is a competing factor. In VALIDATE II, this value is computed by assigning to each atom an integer equal to the maximal number of hydrogen bonds (possibly zero) that it can establish in water; for example, the oxygen in a carbonyl is supposed to form two hydrogen bonds, while an alcoholic group is supposed to form a maximum of three hydrogen bonds. The capacity is then the sum of the integers.

15. VALIDATE II

For a statistical evaluation of the relationship between the parameters and the binding affinities, the program GOLPE (Baroni et al., 1993) was employed. The training set consisted of 49 experimental protease–inhibitor co-crystallized complexes (42 HIV-1, six HIV-2, and one SIV). A PLS run using two principal components on the full 27×49 (27 parameters and 49 examples) data matrix yielded an r^2 of 0.75 and s of 0.74 log units. A leave-one out cross-validation (LOO-CV) produced $q^2 = 0.69$ with $s_{press} = 0.87$ using 2 components, while cross-validation using two groups (LHO-CV) yielded $q^2 = 0.57$. Variable selection by fractional factorial design implemented in GOLPE raised the q^2 to 0.69 on a data matrix with just 8 parameters, with $r^2 = 0.77$ and $s = 0.71$. A more refined hybrid QSAR was obtained balancing the information for each pK_i interval using the same number of compounds for each unit step. A subsequent PLS/LHO-CV lead to a final model showing $r^2 = 0.82$, $s = 0.67$, $q^2 = 0.73$, and $s_{press} = 0.81$ for a data matrix of 9×42. For comparison, the same procedure using only the original 14 VALIDATE parameters did not afford any acceptable HIV-QSAR model ($r^2 = 0.57$, $q^2 = 0.45$).

This model revealed good predictiveness when applied to the test sets (398 compounds total, fig. 3.3) yielding an average absolute global error of prediction (AAE) of 1.03 log units and an r^2_{pred} of 0.46. It should be emphasized that test compounds were simply fitted to the closest template in the training set and then minimized in the active site. No effort to determine the optimal binding mode by rotation of ligand groups or enzyme side chains was performed. On compounds with affinities within the range of the training set, the average absolute error was only 0.89 log units, emphasizing that interpolation within the training set was more accurate.

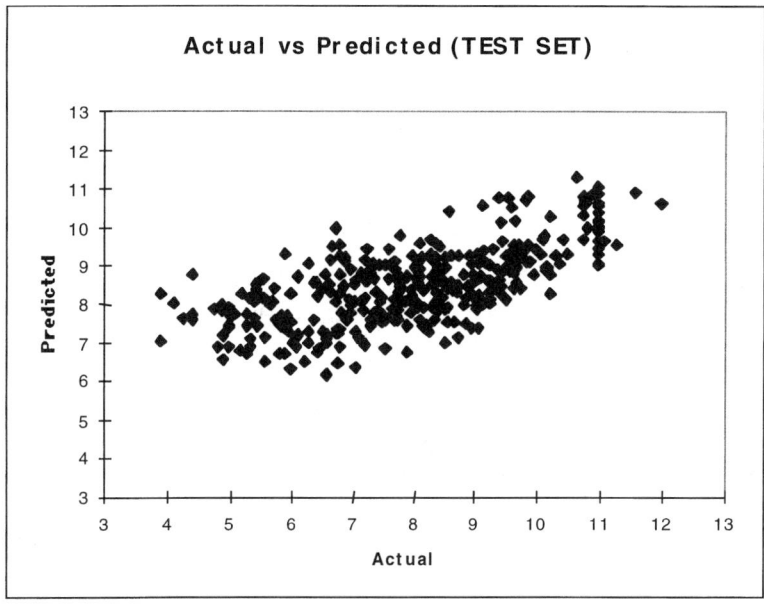

Figure 3.3. Comparison of predicted and observed binding affinities for 398 inhibitors of HIV-1 protease cited in the literature (Ragno and Marshall, unpublished).

Thus, one can use known binding modes as a first approximation to the binding mode of homologous compounds and get good predictive results. Second, one can capture enough of the relevant physical chemistry to predict the activity of compounds that deviate greatly from any of those in the training set. One test set was derived from 3D-database searching, and attempted to position functional groups in the correct orientation (Wang et al., 1996) corresponding to the pharmacophoric groups specified in the database search. From the lack of success in prediction of the affinities of several of these compounds, it is likely that they have a binding mode different from that originally suggested (Wang et al., 1996) from the pharmacophoric pattern.

16. Summary

Approaches to prediction of affinity fall into two camps. The first attempts the more rigorous way through molecular dynamics simulations to accurately predict the free energy of complex formation when the structure and affinity of a complex of an analogous ligand is known. While this process of calculating $\Delta\Delta G$ using the thermodynamic cycle has been shown to be correct for several cases in which the ligand and its congener are quite similar in structure, sampling issues present logistical problems if the structures deviate significantly and the computational demands preclude its use in the iterative loop of de novo design. In response to the need for reasonable estimates of affinity in the design of ligands and prediction of the effects of mutations in protein engineering, a variety of heuristically based methods have been developed. The most accurate of these procedures can predict the affinity of complexes within 1 kcal/mol on average. For a more comprehensive coverage of the complexities involved in predicting affinities, and of the various approaches being tried, see the review by Ajay and Murcko (1995) and the references therein.

Acknowledgments The authors thank the National Institutes of Health for support of research on prediction of affinity and of the time spent on this review. Rino Ragno is the recipient of a fellowship from the Instituto Superiore di Sanita (Rome, Italy). VALIDATE can be downloaded from the Center for Molecular Design's WWW homepage (http://www.cmdsf1.wustl.edu).

References

Abraham, D. J., and Kellogg, G. E. (1993) in: *3D QSAR in Drug Design*, H. Kubinyi (ed.). Leiden: ESCOM Science Publishers. Pp. 506–22.
Ajay and Murcko, M. A. (1995) *J. Medicinal Chem.* **38**, 4593–67.
Åqvist, J., Medina, C., and Samuelson, J.-E. (1994) *Protein Eng.* **7**, 385–91.
Åqvist, J., and Warshel, A. (1989) *Biochemistry* **28**, 4680–9.
Baroni, M., Costantino, G., Cruciani, G., Riganelli, D., Valigi, R., and Clementi, S. (1993) *Quant. Struct.-Act. Relat.* **12**, 9–20.
Bohm, H.-J. (1992) *J. Comput.-Aided Mol. Design* **6**, 61–78.
Bohm, H.-J. (1994) *J. Comput.-Aided Mol. Design* **8**, 243–56.
Collins, J. R., S. K., and Erickson, J. W. (1995) *Nature Structural Biology* **2**, 334–8.
Colman, P. M., Laver, W. G., Varghese, J. N., Baker, A. T., Tulloch, P. A., Air, G. M., and Webster, R. G. (1987) *Nature* **326**, 358–63.
Cramer, R. D., III, Patterson, D. E., and Bunce, J. D. (1988) *J. Am. Chem. Soc.* **110**, 5959–67.
Cummins, P. L., and Gready, J. E. (1993a) *J. Comput.-Aided Mol. Design* **7**, 535–55.
Cummins, P. L., and Gready, J. E. (1993b) *Proteins* **15**, 426–35.
DeWitte, R. S., and Shakhnovich, E. I. (1996) *J. Am. Chem. Soc.* **118**, 11733–44.
Doig, A. J., and Williams, D. H. (1992) *J. Am. Chem. Soc.* **114**, 338–43.
Doweyko, A. M. (1988) *J. Medicinal Chem.* **31**, 1396–406.

Eldridge, M. D., Murray, C. W., Auton, T. A., Paolini, G. V., and Mee, R. P. (1997) *J. Comput.-Aided Mol. Design* **11**.
Ferguson, D. M., Radmer, R. J., and Kollman, P. A. (1991) *J. Medicinal Chem.* **34**, 2654–9.
Fisanick, W., Cross, K. P., and Rusinko, A. (1990) *Tetrahedron Computer Methodology* **3**, 635–52.
Folkers, G., Merz, A., and Rognan, D. (1993) in: *3D-QSAR in Drug Design*, H (ed.). Leiden: Kubinyi ESCOM.) pp. 583–618.
Gilson, M. K., Given, J. A., Bush, B. A., and McCammon, J. A. (1997) *Biophys. J.* **72**, 1047–69.
Go, N., and Scheraga, H. A. (1970) *Macromolecules* **3**, 178–87.
Gorse, A.-D., and Gready, J. E. (1997) *Protein Eng.* **10**, 23–30.
Graves, B. J., Hatada, M. H., Miller, J. K., Graves, M. C., Roy, S., Cook, C. M., Krohn, A., Martin, J. A., and Roberts, N. A. (1991) *Adv. Exptl. Med. Biol.* **306**, 455–60.
Green, S. M., and Marshall, G. R. (1995) *Trends Pharmacol. Sci.* **16**, 285–91.
Hansch, C., and Leo, A. (1979) *Substituent Constants for Correlation Analysis in Chemistry and Biology*. New York: Wiley.
Hansson, T., Marelius, J., and Åqvist, J. (1998) *J. Comput.-Aided Mol. Des.* **12**, 27–35.
Head R. D., Smythe, M. L., Oprea, T. I. Waller, C. L., Green, S. M., and Marshall, G. R. (1996) *J. Am. Chem. Soc.* **118**, 3959–69.
Helms, V., and Wade, R. C. (1998) *J. Am. Chem. Soc.* **120**, 2710–3.
Hermans, J., and Wang, L. (1997) *J. Am. Chem. Soc.* **119**, 2707–14.
Ho, C. M. W., and Marshall, G. R. (1994) in: *Proceedings of the twenty-seventh annual Hawaii International Conference on system sciences*, L. Hunter (ed.). Washington, DC: IEEE Computer Society Press. Vol. 5, pp. 212–22.
Holloway, M. K., Wai, J. M., Halgre, T. A., Fitzgerald, P. M. D., Vacca, J. P., Dorsey, B. D., Levin, R. B., Thompson, W. J., Chen, L. J., deSolms, S. J., Gaffin, N., Ghosh, A. K., Giuliani, E. A., Graham, S. L., Guare, J. P., Hungate, R. W., Lyle, T. A. Sanders, W. M., Tucker, T. J., Wiggins, M., Wiscount, C. M., Woltersdorf, O. W., Young, S. D., Darke, P. L., and Zugay, J. A. (1995) *J. Medicinal Chem.* **38**, 305–17.
Horton, N., and Lewis, M. (1992) *Protein Sci.* **1**, 169–81.
Hulten, J., Bonham, N. M., Nillroth, U., Hansson, T., Zuccarello, G., Bouzide, A., Åqvist, J., Classon, B., Danielson, U. H., Karlen, A., Kvarnstrom, I., Samuelsson, B., and Hallberg, A. (1997) *J. Medicinal Chem.* **40**, 885–97.
Jackson, R. M., and Sternberg, M. J. (1995) *J. Mol. Biol.* **250**, 258–75.
Jorgensen, W. L., and Ravimohan, C. (1985) *J. Chem. Phys.* **83**, 3050–4.
Kellogg, G. E., Semus, S. E., and Abraham, D. J. (1991) *J. Comput. Aided Mol. Des.* **5**, 545–52.

Keske, J., and Dixon, J. S. (1995) In: *New Methods in Drug Research*, A. Makriyannis and N. Castignoli (eds.). Barcelona: J. R. Prous Science Publishers.

Kollman, P. (1993) *Chem. Rev.* **93**, 2395–417.

Krystek, S. R., Jr., Bruccoleri, R. E. and Novotny, J. (1991) *Int. J. Pept. Protein Res.* **38**, 229–36.

Kubinyi, H. (1993) (ed.) *3D QSAR in Drug Design*. Leiden: ESCOM Science Publishers. pp. 1–759.

Le Grand, S. M., and Merz, K. M., Jr. (1993) *J. Comput. Chem.* **14**, 349–52.

Marelius, J., Graffner-Nordberg, M., Hansson, T., Halberg, A., and Åqvist, J. (1998) *J. Comput.-Aided Mol. Design* **12**, 1–13.

Marshall, G. R., and Cramer R. D., III (1988) *Trends Pharmacol. Sci.* **9**, 285–9.

Marshall, G. R., and Ragno, R. unpublished results.

Martin, Y. C. (1978) *Quantitative Drug Design: A Critical Introduction.* New York: Marcel Dekker.

McDonald, I. K., and Thornton, J. M. (1994) *J. Mol. Biol.* **238**, 777–93.

Miranker, A., and Karplus, M. (1991) *PROTEINS: Structure, Function and Genetics* **11**, 29–34.

Mohamadi, F., Richards, N. G. J., Guida, W. C., Liskamp, R., Lipton, M., Caufield, C., Chang, G., Hendrickson, T., and Still, W. C. (1990) *J. Comput. Chem.* **11**, 440–67.

Murphy, K. P., Xie, D., Thompson, K. S., Amzel, L. M., and Freire, E. (1994) *Proteins: Structure, Function and Genetics* **18**, 63–7.

Nishibata, Y., and Itai, A. (1991) *Tetrahedron* **47**, 8985–90.

Novotny, J., Bruccoleri, R. E., and Saul, F. A. (1989) *Biochemistry* **28**, 4735–49.

Oprea, T., I., Waller, C. L., and Marshall, G. R. (1994a) *Drug Design & Discovery* **12**, 29-51.

Oprea, T. I., Waller, C. L., and Marshall, G. R. (1994b) *J. Medicinal Chem.* **37**, 2206–15.

Ortiz, A. R., Pisabarro, M. T., Gago, E, and Wade, R. (1995) *J. Medicinal Chem.* **38**, 2681–91.

Paulsen, M. D., and Ornsten, R. L. (1997) *Protein Eng.* **9**, 567–71.

Pearlman, D. A., and Murcko, M. A. (1993) *J. Comput. Chem.* **14**, 1184–93.

Pickett, S. D., and Sternberg, M. J. (1993) *J. Mol. Biol.* **231**, 825–39.

Radmer, R. J., and Kollman, P. J. (1998) *J. Comput.-Aided Mol. Des.* **14**, 1–13.

Rao, B. G., Tilton, R. E., and Singh, U. C. (1992) *J. Am. Chem. Soc.* **114**, 4447–52.

Reddy, M. R., Viswanadhan, V. N., and Weinsten, J. N. (1991) *Proc.Natl. Acad. Sci. USA* **88**, 10287–91.

Rich, D. H., Sun, C.-Q., Vara Prasad, J. V. N., Pathiasseril, A., Toth, M. V., Marshall, G. R., Clare, M., Mueller, R. A., and Houseman, K. (1991) *J. Med. Chem.* **34**, 1222–5.

Rotstein, S. H., and Murcko, M. A. (1993) *J. Medicinal Chem.* **36**, 1700–10.
Searle, M. S., Westwell, M. S., and Williams, D. H. (1995) *J. Chem. Soc.* [Perkin 2], 141–51.
Searle, M. S., and Williams, D. H. (1992) *J. Am. Chem. Soc.* **114**, 10690–7.
Searle, M. S., Williams, D. H., and Gerhard, U. (1992) *J. Am. Chem. Soc.* **114**, 10697–704.
Still, W. C., Tempczyk, A., Hawley, R. C., and Hendrickson, T. (1990) *J. Am. Chem. Soc.* **112**, 6127–9.
Straatsma, T. P., and McCammon, J. A. (1991) *Methods in Enzymology* **202**, 497–511.
Tropsha, A., and Hermans, J. (1992) *Protein Eng.* **5**, 29–33.
Vajda, S., Weng, Z., Rosenfeld, R., and DeLisi, C. (1994) *Biochemistry* **33**, 13977–88.
Waller, C. L., and Marshall, G. R. (1993) *J. Medicinal Chem.* **36**, 2390–403.
Waller, C. L., Oprea, T. I., Giolitti, A., and Marshall, G. R. (1993) *J. Medicinal Chem.* **36**, 4152–60.
Wang, S., Milne, G. W. A., Yan, X., Posey, I. J., Nicklaus, M. C., Graham, L., and Rice, W. G. (1996) *J. Medicinal Chem.* **39**, 2047–54.
Weng, Z., Delisi, C., and Vajda, S. (1997) *Protein Sci.* **6**, 1976–84.
Williams, D. H. (1991) *Aldrichimica Acta* **24**, 71–80.
Williams, D. H., Cox, J. P. L., Doig, A. J:, Gardner, M., Grehard, U., Kaye, P. T., Lal, A. R., Nicholls, I. A., Salter, C. J., and Mitchell, R. C. (1991) *J. Am. Chem. Soc.* **113**, 7020–30.
Wold, S., Albano, C., Dunn, W. J., III, Esbensen, K., Hellberg, S., Johansson, E., Lindberg, W., and Sjostrom, M. (1984) *Analysis* **12**, 477–85.
Wold, S., Johansson, E., and Cocchi, M. (1993) In: *3D QSAR in Drug Design*, H. (ed.). Leiden: Kubinyi ESCOM Science Publishers. Pp. 523–50.

4

Electrostatic Interactions in Proteins and Nucleic Acids: Theory and Applications

Kim A. Sharp

Electrostatic interactions play a crucial role in the determining and maintaining the structures of proteins and nucleic acids. Electrostatic interactions also affect the reactivity and functional properties of proteins and nucleic acids when movement of charged or dipolar groups and molecules are involved. Examples include charge-transfer processes such as protonation, ion binding, and electron transfer; redistribution of charge during catalytic reactions; and association and binding of charged ligands.

A theoretical treatment of electrostatic interactions in macromolecules has three components: (1) a physical model for describing the response of the system to the electrostatic field arising from the macromolecular charge, (2) A numerical method for calculating the resulting electrostatic potential distribution, and (3) the relationship between the electrostatic potential distribution and the macromolecular property or experimental observable of interest. Physical models and numerical methods for treating macromolecular electrostatics have been reviewed in several publications and will be described only briefly. The relationship between the electrostatic potential and the macromolecular properties will be the focus of this chapter. Since the movement of a charge in an electrostatic field results in thermodynamic work, this requires a combination of electrostatics and thermodynamic or statistical-mechanical theory. The Poisson–Boltzmann model for electrostatics will be used throughout to illustrate how one may calculate these electrostatic interactions. It should, however, be stressed that other models for electrostatics can, and are, used to treat electrostatic interactions in proteins and nucleic acids.

1. Theory

1.1. The Molecular Charge Distribution

The first step in calculating the electrostatic potential distribution of a macromolecule is to describe its charge distribution. This may represent either the instantaneous or average state and structure of the molecule and its surrounding environment and solvent, depending on the application. In most situations of interest, using a complete description of the charge state and motions of all atoms is neither possible or desirable. Thus some charges are represented explicitly, while either the others are determined within the calculation itself (so-called induced or mobile charges), or their electrostatic *effect* is included implicitly. Usually the charges on the molecule itself are represented explicitly, while solvent and solvent-ion charges are represented implicitly.

The explicit charge distribution requires the molecular coordinates, typically obtained from x-ray crystallography, NMR, or molecular mechanics, and the charge assigned to each atom. Various atomic charge sets suitable for classical electrostatic calculations are available from quantum-mechanical calculations and/or parametrization to experimental data.

1.2. The Response of the System to Electrostatic Fields

Given a certain explicit macromolecular charge distribution the electrostatic potential distribution depends on how the system (molecule plus surrounding solvent) responds to the field produced by that charge distribution. The response arises from three physical processes: Electronic polarization, reorientation of permanent dipolar groups, and movement of whole charges. The latter usually involves redistribution of mobile ions in the solvent, although it may include movement of ionized side chains if significant changes in protein structure occur.

Electronic polarizability can be represented by point inducible dipoles or by a dielectric constant. For the latter, the polarization P (the mean dipole moment per unit volume at position r) is proportional to the Maxwell (total) field E, and the dielectric constant, ε:

$$P(r) = (\varepsilon(r) - 1)E(r)/4\pi. \qquad (1)$$

The polarization is not usually calculated directly using eq. (1), since E itself depends on P. Instead the potential and polarization distributions are calculated self-consistently using, for example, the Poisson equation (described below). The electronic polarizability of water and most organic material, evaluated by high-frequency dielectric (ε^∞) measurements or the refractive index, is very similar, and so can be well represented by a single dielectric constant of about 2.

The reorientation of groups that have large permanent dipoles, such as peptides or water molecules, is an important part of the dielectric response. This too may be treated using a dielectric constant: using eq. (1) with a dielectric constant greater than 2. Four factors determine the dielectric contribution from permanent dipoles: (i) the dipole-moment magnitude, (ii) the density of such groups in the molecule or solvent, (iii) the freedom of such groups to reorient, and (iv) the degree of cooperativity between dipole motions. Factors i–iv are all large for water, so consequently it has a high dielectric constant (78.6 at 25°C)). Although the first two factors are also large for macromolecules like proteins and nucleic acids (so they are very polar), they usually have a defined structure, so the freedom of dipolar groups to reorient is much smaller than in water, and consequently their dielectric (or polarizability) is significantly smaller. Calculation of isotropic dielectric constants for amorphous protein solids (Gilson and Honig, 1986; Nakamura et al., 1988) and the interior of proteins in solution (Simonson and Perahia, 1995; Smith et al., 1993) and experimental measurements (Takashima and Schwan, 1965) provide an estimate of $\varepsilon = 2.5$–4.0 for the contribution of dipolar groups to the protein dielectric.

The dielectric model can also be extended to incorporate saturation effects (Warwicker, 1994), although there is a compensation effect of electrostriction, which increases the local dipole density (Jayaram et al., 1989a), so that the linear solvent dielectric model has proven sufficiently accurate for most protein and nucleic-acid applications to date.

Charge rearrangement can occur in any circumstance where there are formally charged groups that have some degree of mobility. The most mobile, and easiest to treat theoretically, are salt ions in the solvent. The commonest way of treating this is via the Boltzmann model, where the charge density is given by

$$\rho^m(\mathbf{r}) = \sum_i z_i e c_i^0 e^{-z_i e \phi(\mathbf{r})/kT}, \tag{2}$$

where c_i^0 is the bulk concentration of the ion of type i, and of valence z_i, and $\phi(\mathbf{r})$ is the average potential at $\mathbf{r}$. Thus cations will be found preferentially in regions of negative potential around a molecule, and depleted in regions of positive potential, and vice versa for anions. Equation (2) neglects the effect of ion size and correlation between ion positions, but comparison with more detailed mobile ion treatments shows it to be quite accurate for monovalent (1-1) salts at mid-range concentrations of 0.001–0.5 M (Bacquet and Rossky, 1984; Murthy et al., 1985; Olmsted et al., 1989; Olmsted et al., 1991; Record et al., 1990; Sharp, 1995).

Another form of response comes from formally charged groups on macromolecules, particularly the longer side chains on the surface of proteins—for example, arginine and lysine. Two approaches to modeling these formal charge movements can be taken. The first is to treat them as part of the dielectric response. In this case, molecular-dynamic (MD) simulations of the dipole fluctuations arising from charge movements suggest a dielectric in the range of 20–30 in the regions of these charged groups (Simonson and Perahia, 1995; Smith et al., 1993). A better approach is to model the effect of charge motions on the electrostatic quantity of interest explicitly with molecular-mechanical (MM) simulations, without formulating these motions as a dielectric at all (Langsetmo et al., 1991; Wendoloski and Matthew, 1989). This may require generating an ensemble of structures with different explicit charge distributions. Alternatively, one is often interested in a specific biological process A → B, in which one can evaluate the structure of the protein in states A and B (experimentally or by modeling), and any change in average charge positions is incorporated at the level of different average explicit charge-distribution inputs for the calculation, modeling only the electronic, dipolar, and salt contributions as the response.

1.3. Dependence of the Potential on the Charge Distribution

The potential at a point in space, r, arising from one distribution of charge $\rho(s)$ and polarization $P(s)$ is given by

$$\phi(r) = \int_V \left(\frac{\rho(s)}{|(s-r)|} + \frac{P(s) \cdot (s-r)}{|s-r|^3} \right) ds. \tag{3}$$

The total charge distribution is the sum of the explicit charge distribution on the molecule and that from the mobile solvent-ion distribution, $\rho = \rho^e + \rho^m$. Substituting for the dielectric polarization using eq. (1), and the mobile-ion charge distribution using eq. (2), we can express the potential in terms of a partial differential equation, the Poisson–Boltzmann (PB) equation

$$\nabla \cdot \varepsilon(r)\nabla\phi(r) + 4\pi \sum_i z_i e c_i^o e^{-z_i e\phi(r)/kT} + 4\pi\rho^e(r) = 0, \tag{4}$$

which relates the potential, molecular-charge, and dielectric distributions, $\phi(r)$, $\rho^e(r)$, and $\varepsilon(r)$, respectively. Contributions to the polarizability from electrons, from the molecule's permanent dipoles, and from solvent dipoles are incorporated into this model by using an appropriate value for the dielectric for each region of protein and solvent. The potential at a particular charged atom i thus includes three physically distinct contributions. The first is the direct or Coulombic potential of j at i. The second is the potential at i from the polarization (of the molecule, water, and ion atmosphere) induced by j. This is the screening potential which opposes the direct Coulombic potential. The third contribution arises from the polarization induced by i itself. This is the reaction, solvation, or self potential. This potential is of opposite sign to the charge on i, thus preferentially stabilizing, or solvating, the charge in a more polarizable environment. The direct, screening, and reaction potential contributions may be calculated separately if required by appropriate use of the PB equation. For example, the self potential is obtained from calculations in which charge is assigned only to group i, while the screening contribution to the i–j interaction can be obtained by taking the difference between values obtained for the potential at i in vacuo and with solvent with charge assigned to j only.

For protein applications, the Boltzmann term in eq. (4) is usually linearized to become $-8\pi\phi(\mathbf{r})I/kT$, where I is the ionic strength, whereas for nucleic acids, and molecules of similarly high charge density, the full nonlinear equation is used.

1.4. Calculation of Energies and Forces

Once the electrostatic potential distribution has been obtained, calculation of experimental properties usually requires evaluation of an electrostatic energy or force. For a linear system (where the dielectric and ionic responses are linear), the electrostatic free energy is given by

$$\Delta G^{\text{el}} = \tfrac{1}{2} \sum_i \phi_i q_i, \qquad (5)$$

where ϕ_i is the potential at an atom with charge q_i. The most common source of nonlinearity is the Boltzmann term in the PB equation (4) for highly charged molecules such as nucleic acids. The total electrostatic energy in this case is (Reiner and Radke, 1990; Sharp and Honig, 1990; Zhou, 1994)

$$\Delta G^{\text{el}} = \int_V \left(\rho^e \phi - \frac{\varepsilon E^2}{8\pi} - kT \sum_i c_i^o (e^{-z_i e\phi/kT} - 1) \right) d\mathbf{r}, \qquad (6)$$

where the integration is taken over the molecular and solvent volumes. The electrostatic energy evaluated using dielectric models is a free energy, since it represents thermodynamic work done by transferring charge. Furthermore, a consideration of the physical behavior behind the dielectric constant of, for example, water shows that the degree to which the dipoles are aligned by an applied field depends on a balance between electrostatic forces and the unfavorable entropy of dipole ordering. Consequently, the dielectric constant is temperature-dependent, which will impart a temperature dependence, and hence an entropic component, to the free energy. Examination of the temperature dependence of the electrostatic free energy allows one to analyze entropic, enthalpic, and heat-capacity components of electrostatic free energies with these dielectric models (Gallagher and Sharp, 1997; Rashin and Honig, 1985; Sharp, 1995).

The general expression for the electrostatic force on a charge q is given by the gradient of the total free energy with respect to that charge's position

$$\boldsymbol{f}_q = -\nabla_{r_q}(G^{\text{el}}). \tag{7}$$

If the movement of that charge does not affect the potential distribution, due to the other charges and dipoles, then eq. (7) can be evaluated using the 'test-charge' approach, in which case the force depends only on the gradient of the potential, or the field, at the charge:

$$\boldsymbol{f} = q\boldsymbol{E}. \tag{8}$$

The test-charge approach is not always applicable, however (Davis and McCammon, 1990). For example, forces arise between a single charge and a dielectric boundary due to image-charge (reaction-potential) effects. Also, movement of charged atoms may further result in movements of the dielectric boundary between the solute and solvent, which will change the energy of the system. A similar effect to the 'dielectric-pressure' forces arises from solvent-ion pressure at the solute–solvent boundary. These forces generally act to increase the interaction of polar or charged atoms with the higher-dielectric solvent, by moving them towards the molecular surface, and into the solvent. A general expression for the force density that includes both these effects has been derived within the PB model using variational methods (Gilson et al., 1993):

$$\boldsymbol{f} = \rho^e \boldsymbol{E} - \tfrac{1}{2}E^2 \nabla \varepsilon - kT \sum_i c_i^0 (e^{-z_i e\phi/kT} - 1)\nabla A, \tag{9}$$

where A is a function describing the accessibility to solvent ions, which is 0 inside the protein, and 1 in the solvent, and whose gradient is nonzero only at the solute–solvent surface. Similarly, in a two-dielectric model (solvent plus molecule), the gradient of ε is nonzero only at the molecular surface. The first term accounts for the force acting on a charge due to a field, as in eq. (8), while the second and third terms account for the dielectric-surface and ionic-atmosphere pressure terms respectively. Equation (9) has been used to combine the PB equation and molecular mechanics (Gilson et al., 1995).

1.5. Numerical Methods

In most current models of macromolecular electrostatics, the molecule is represented by the positions, charges, and radii of all the atoms which describe its shape and polarity. The region inside the molecule is assigned a low dielectric, which accounts for electronic polarizability and, if appropriate, dipolar polarizability. The region outside the molecular surface is assigned the solvent dielectric value and the required ionic strength. To obtain the electrostatic potential requires the ability to solve the equations for this complex asymmetric distribution of charge, dielectric, and salt concentration. A variety of numerical methods have been applied to the PB equation. Inevitably, this involves discretizing space and using finite methods. These include finite-difference methods (Gilson et al., 1998; Nicholls and Honig, 1991; Warwicker and Watson, 1982), finite-element methods (Rashin, 1990; Yoon and Lenhoff, 1992; Zauhar and Morgan, 1985), multigridding (Holst and Saied, 1993; Oberoi and Allewell, 1993), conjugate-gradient methods (Davis and McCammon, 1989), and fast multipole methods (Bharadwaj et al., 1994; Davis, 1994). Methods for treating the nonlinear PB include under-relaxation (Jayaram et al., 1989b) and powerful inexact Newton methods (Holst et al., 1994a). The nonlinear PB equation can also be solved via a self-consistent-field approach, in which one calculates the potential using eq.(3); then the polarization and mobile charge density are calculated using eqs. (1) and (2), respectively, and the procedure is repeated until convergence (Pack and Klein, 1984; Pack et al., 1986). The method allows one to put in more-elaborate models for the ion distribution, for example incorporating the finite size of the ions (Pack et al., 1993). Approximate methods based on spherical approximations (Born-type models) have also been used (Schaeffer and Frommel, 1990; Still et al., 1990). Considerable progress has been made in finite methods, and accurate rapid algorithms are available. The reader is referred to the original references for numerical details. Values for protein atomic charges, radii, and dielectric constants suitable for use with the Poisson–Boltzmann equation are available in the literature (Jean-Charles et al., 1990; Mohan et al., 1992; Simonson and Brunger, 1994; Sitkoff et al., 1994).

2. Applications

2.1. Solvation Energies

The solvation energy is defined as the change in free energy upon transfer of a molecule or group from one solvent (or environment) to another. Solvation energy changes are an important component of shifts in charge-transfer equilibria, of the total free energy of binding, and other processes. Free energies of transfer of small molecules between solvents are also an important source of data for parametrizing and testing electrostatic models. The electrostatic component of the solvation free-energy change is obtained as the difference in reaction field or self energy of a group or molecule in the two environments, A and B

$$\Delta G^{\text{rf}} = \tfrac{1}{2} \sum_i q_i(\phi_i^{\text{rf}}(B) - \phi_i^{\text{rf}}(A)), \tag{10}$$

where the summation is over all charges q_i of the group or molecule, and the potential is that due to the reaction field of those charges. To calculate total solvation energy changes, eq. (1) must be combined with a method for calculating the nonelectrostatic solvation contribution, e.g. by surface-area methods. Applications to solvation include comparisons of electrostatic methods and free-energy perturbation (Ewing and Lybrand, 1993; Jean-Charles et al., 1990) and analysis of small-molecule and ion solvation (Mohan et al., 1992; Rashin and Namboodiri, 1987; Sharp et al., 1992; Simonson and Brunger, 1994; Sitkoff et al., 1994).

2.2. Charge-transfer Equilibria

Many functions of proteins and nucleic acids require the transfer of charge. Important examples include transfer of protons, electrons, and ions, which are governed by acid–based equilibria, electron-transfer redox midpoints, and ion-binding equilibria respectively. The theory of the dependence of these three equilibria on classical electrostatic effects can be treated in an identical manner.

Beginning with acid-base equilibria, a titratable group will have an intrinsic ionization equilibrium, expressed in terms of a known intrinsic $\text{p}K_a^o$ (where $\text{p}K_a^o = -\log_{10}(K_a^o)$, and K_a^o is the dissociation constant for the reaction $H^+A = H^+ + A$, with A an acid or a base) which is determined by all the quantum-chemical, electrostatic, and environmental effects operating on that group in some reference state. For example, a reference state for the ionization of the side chain of aspartic acid might be the isolated amino acid in water, for which $\text{p}K_a^o = 3.85$. In the environment of a protein, the $\text{p}K_a$ will be altered by three electrostatic effects. The first is from the different polarizability of the environment, the second is due to interaction with permanent dipoles in the protein, and the third is due to other charged—perhaps titratable or redox—groups. The effective $\text{p}K_a$ is given by

$$\text{p}K_a = \text{p}K_a^o + (\Delta\Delta G^{\text{rf}} + \Delta\Delta G^{\text{perm}} + \Delta\Delta G^{\text{tit}})/2.303kT, \tag{11}$$

where the factor of $1/2.303kT$ converts units of energy to units of $\text{p}K_a$. The first contribution, $\Delta\Delta G^{\text{rf}}$, arises because the completely solvated group induces a strong favorable reaction field in the high-dielectric water, which stabilizes the charged form of the group (the neutral form is also stabilized by the solvent reaction field induced by any dipolar groups, but to a lesser extent). Desolvating the group to any degree by moving it into a less polarizable environment will preferentially destabilize the charged form of that group, shifting the pK_a by an amount

$$\Delta\Delta G^{\text{rf}} = \tfrac{1}{2}\sum_i \left(q_i^{\text{d}}\Delta\phi_i^{\text{rf,d}} - q_i^{\text{p}}\Delta\phi_i^{\text{rf,p}}\right), \tag{12}$$

where q_i^{p} and q_i^{d} are the charge distributions on the group, $\Delta\phi_i^{\text{rf,p}}$ and $\Delta\phi_i^{\text{rf,d}}$ are the changes in the group's reaction potential upon moving it from its reference state into the protein, in the protonated (superscript p) and deprotonated (superscript d) forms, respectively, and the sum is over the group's charges. The contribution of the permanent dipoles is given by

$$\Delta\Delta G^{\text{perm}} = \sum_i \left(q_i^{\text{d}} - q_i^{\text{p}}\right)\phi_i^{\text{perm}}, \tag{13}$$

where ϕ_i^{perm} is the interaction potential at the ith charge due to all the permanent dipoles in the protein, including the effect of screening. Intrinsic pK_a values of groups in proteins are rarely shifted by more than 1 pK_a unit, indicating that the effects of desolvation are often compensated to a large degree by the $\Delta\Delta G^{\text{perm}}$ term.

The final term accounts for the contribution of all the other charge groups:

$$\Delta\Delta G^{\text{tit}} = \sum_i \left(q_i^{\text{d}}\langle\phi_i\rangle_{\text{pH},c,\Delta V}^{\text{d}} - q_i^{\text{p}}\langle\phi_i\rangle_{\text{pH},c,\Delta V}^{\text{p}}\right), \tag{14}$$

where $\langle\phi_i\rangle$ is the mean potential at group charge i from all the other titratable groups. The charge state of the other groups in the protein depend in turn on their intrinsic pK_a values, on the external pH if they are acid–based groups, on the external redox potential ΔV if they are redox groups, and on the concentration c of ions if they are ion-binding sites, as indicated by the subscript on $\langle\phi_i\rangle$. Moreover, the charge state of the group itself will affect the equilibrium at the other sites. Because of this linkage, exact determination of the complete charged state of a protein is a complex procedure. If there are N such groups, the rigorous way to do this is to compute the titration-state partition function by evaluating the relative electrostatic free energies of all 2^N ionization states for a given set of pH, c, and ΔV values. From this, one may calculate the mean ionization state of any group as a function of pH, ΔV, and so on. For large N, this becomes impracticable, but various approximate schemes work well, including a Monte Carlo procedure (Beroza et al., 1991; Yang et al., 1993) or partial evaluation of the titration partition function by clustering the groups into strongly interacting subdomains (Bashford and Karplus, 1990; Gilson, 1993; Yang et al., 1993).

Calculation of ion-binding equilibria in proteins proceeds exactly as for calculation of acid–based equilibria, the results usually being expressed in terms of changes in an associated constant, K_a for the reaction $Y^z + A = Y^z A$, where z is the ion valence. Similarly for shifts in the redox midpoint potential, defined by [oxidized]/[reduced]

= $\Delta V - E_m$, where the midpoint potential E_m is the external reducing potential at which the group is half oxidized and half reduced. Shifts in E_m can be calculated using the equivalent of eq. (11), where now the conversion factor of $25.7/kT$ converts the electrostatic energies to the customary units of mV. E_m values are customarily measured and tabulated at pH 7 (E_{m7}), since redox midpoints, like pK_a values (and ion binding equilibria) are dependent on pH, ion concentration, and the protein's charged state.

Shifts in charge-transfer equilibria have a number of consequences. Using the proton-transfer event as an example, consider some conformational change or binding event (etc.) X → Y that shifts the pK_a of a group i. If the pK_a is shifted to or from a value near the pH, this event will result in release or uptake of protons. Calculation of this release is straightforward once all the pK_a values are known. A reciprocal consequence is that the process X → Y will have a pH dependence. Furthermore, removal of group i or replacement by one with a different pK_a will affect the energetics of X → Y. If we have pK_a^x and pK_a^y in the two states X and Y respectively, the release of protons is given by

$$\Delta n_{H^+} = \frac{1}{1 + 10^{pH-pK_a^x}} - \frac{1}{1 + 10^{pH-pK_a^y}}, \tag{15}$$

which has a maximum at a pH midway between the two pK_a values. The equilibrium between X and Y, given by $K = [X]/[Y]$, is

$$K/K_o = \frac{1 + 10^{-(pH-pK_a^x)}}{1 + 10^{-(pH-pK_a^y)}}, \tag{16}$$

where K_o is the high-pH limiting value of the equilibrium constant (Alternatively the low-pH state may be used as a refererence). The change in free energy for the process X → Y is given by $-kT \ln(K/K_o)$. Equations (15–16) can be generalized to deal with a collection of linked groups. Analogous considerations apply to the release of bound ions or electrons, and to the dependence of some equilibria on salt concentration or redox potential.

There have been many studies of the electrostatics of charge-transfer equilibria, which include the effect of mutations on active-site histidine pK_a values and enzyme activity in subtilisin (Gilson and Honig, 1987; Sternberg et al., 1987); the linkage between redox potentials and pK_a values in azurin (Bashford et al., 1988); the relationship between pK_a values, salt bridges, and stability in thioredoxin (Langsetmo et al., 1991), and between pK_a values and stability in myoglobin (Yang and Honig, 1994); a detailed analysis of redox potential shifts in ion-sulfur proteins (Langen et al., 1992) and cytochromes (Gunner and Honig, 1990, 1991); and proton relay events in bacteriorhodopsin (Bashford and Gerwert, 1992; Sampogna and Honig, 1994).

2.3. Electron-transfer Reorganization Energies

About a third of all known enzymes are redox proteins, i.e. they transfer electrons as part of their function. The key factors controlling the rate of electron transfer, k^e, are the electronic coupling term, H_{ab}, which describes the overlap of the donor and acceptor orbital systems, and the free energy of activation, $\Delta G^{\pm}$, where

$$k^e = \frac{4\pi^2 H_{ab}^2}{h} e^{-\delta G^{\pm}/kT}, \qquad (17)$$

in which k and h are Boltzmann's and Planck's constants, respectively, and T is the temperature. For weakly coupled systems, the activation energy is given by (Marcus, 1956)

$$\Delta G^{\pm} = (\Delta G^o + \lambda)^2/4\lambda, \qquad (18)$$

where ΔG^o is the driving force, or different in redox midpoints of the donor and acceptor, and λ is the reorganization energy. The physical meaning of λ may be obtained by considering a (hypothetical) pathway in which the electron is first transferred rapidly ($t < 1$ fs) to the acceptor, so that there is electronic relaxation, but no relaxation of the nuclear configurations from their equilibrium donor state. This is followed by relaxation of the nuclear coordinates to the equilibrium acceptor state. Then λ is the energy released upon relaxation; it is thus a kind of reaction field energy. Although λ is a nonequilibrium property, it is possible to calculate it with the Poisson–Boltzmann electrostatic model using (Sharp, 1997)

$$\lambda = \tfrac{1}{2}\sum_i (\delta\phi_i^{d*d} - \delta\phi_i^{ad})\delta\rho_i^{ad}, \qquad (19)$$

where $\delta\rho^{ad}$ is the change in atomic charge distribution upon electron transfer; here, $\delta\phi^{ad}$ is the accompanying change in potential at equilibrium, i.e. allowing full relaxation of the electronic, dipolar, and ionic responses to the electron movement, while $\delta\phi^{d*d}$ is the corresponding change in potential due to reequilibration of electronic polarization only (i.e. assigning a dielectric of ≈ 2 throughout). The sum is over all atoms of the donor and acceptor whose charge changes upon electron transfer. This approach has been used to calculate λ for intramolecular electron-transfer reactions in the photosynthetic reaction center from *R. viridis* and the ruthenated heme proteins cytochrome *c*, myoglobin, cytochrome *b*, and for intermolecular electron transfer between two cytochrome *c* molecules (Sharp, 1997). The overall agreement with experiment is good, and the calculations show that acceptor/donor separation and position of the cofactors with respect to the protein/solvent boundary are equally important and, along with the overall polarizability of the protein, are the major determinants of λ.

2.4. Electrostatic Contributions to Binding Energy

The electrostatic contribution to the binding energy of two molecules is obtained by taking the difference in total electrostatic energies in the bound (AB) and unbound (A + B) states. For the linear case, using eq. (5) gives

$$\Delta\Delta G_{\text{bind}}^{\text{elec}} = \tfrac{1}{2}\sum_i^{N_A} q_i(\phi_i^{AB} - \phi_i^{A}) + \tfrac{1}{2}\sum_j^{N_b} q_j(\phi_j^{AB} - \phi_j^{B}), \quad (20)$$

where the first and second sums are over all charges in molecule A and B, respectively, and ϕ^x is the total potential produced by x =A, B, or AB. From this equation, it should be noted that the electrostatic free-energy change of each molecule has contributions from intermolecular charge–charge interactions, and from changes in the solvent reaction potential of the molecule itself when solvent is displaced by the other molecule. Equation (20) allows for the possibility that the conformation may change upon binding, since different charge positions may be used for the complexed and uncomplexed forms of A, and similarly for B. However other energetic terms, including those involved in any conformational change, have to be added to obtain net binding free-energy changes. Nevertheless, changes in binding free energy due to charge modifications or changes in external factors such as pH and salt concentration may be estimated using eq. (2) alone. Salt effects are usually significant in highly charged molecules, in which case the nonlinear form for the total electrostatic energy, eq. (6), must be used instead of eq. (20). The salt dependence of binding of drugs and proteins binding to DNA has been studied using this approach (Misra et al., 1994a; Misra et al., 1994b; Sharp et al., 1995), including the pH dependence of drug binding (Misra and Honig, 1995). Other applications include the binding of sulfate to the sulfate-binding protein (Åqvist et al., 1991) and antibody and antigen interactions (Lee et al., 1992; Slagle et al., 1994).

2.5. Electrostatic Contributions to Enthalpy, Entropy, and Heat Capacity

Examination of the Poisson–Boltzmann equation (4) shows that a temperature dependence of the free energy comes from two sources: (i) the explicit dependence on temperature in the Boltzmann factor governing the distribution of mobile ions (the second term), (ii) from an implicit dependence via the temperature dependence of the dielectric constant. Since the dielectric response of water involves a large and entropically unfavorable reorientation of water dipoles, it has a strong temperature dependence. This temperature dependence can be incorporated into the calculations by using the appropriate experimentally determined value for the aqueous dielectric constant ($\epsilon_{\text{solvent}}$) at different temperatures (Lide,

1990). This enables estimates of the electrostatic enthalpy, entropy, and heat capacity contributions to be calculated using the PB equation. Taking the temperature derivative using the chain rule gives

$$\Delta S^{\text{elec}} = -\frac{\partial \Delta G^{\text{elec}}}{\partial T} = -\left[\frac{\partial \Delta G^{\text{elec}}}{\partial \epsilon}\right] \frac{\partial \epsilon}{\partial T} - \left[\frac{\partial \Delta G^{\text{elec}}}{\partial T}\right]_{\epsilon}, \quad (21)$$

where the first term is the aqueous dipole-reorientation term (evaluated keeping the Boltzmann factor constant), and the second term is the ion-reorganization term (evaluated by holding the dielectric constant). The same approach can be applied to obtain the enthalpic contributions, using the van't Hoff relation

$$\Delta H = \frac{d}{d(1/T)} \frac{\Delta G}{T}. \quad (22)$$

The derivatives can be evaluated numerically. This approach has been used to evaluate enthalpic contributions to solvation (Rashin and Bukatin, 1993; Rashin and Honig, 1985; Rashin and Namboodiri, 1987), and enthalpic and entropic contributions to salt effects on DNA–ligand binding (Sharp, 1995; Sharp et al., 1995).

This approach can be extended to study the electrostatic contribution to heat capacity by taking the second derivative of the electrostatic free energy with respect to temperature:

$$C_p = -T\left(\frac{\partial^2 G}{\partial T^2}\right)_p. \quad (23)$$

As for the entropy and enthalpy, the complete second derivative of the free energy with respect to temperature (eq. 23) can be broken down into a sum of terms involving the partial derivatives with respect to the explicit (ionic Boltzmann-factor) and implicit (dielectric) dependence on temperature:

$$\Delta C_p = \left[\frac{\partial^2 \Delta G}{\partial T^2}\right]_{\epsilon} + \left[\frac{\partial^2 \Delta G}{\partial \epsilon^2}\right]_{\text{BF}} \left(\frac{\partial \epsilon}{\partial T}\right)^2 + \text{cross terms}. \quad (24)$$

The cross terms involve mixed derivative products of the type $\frac{\partial^2 \Delta G}{\partial T \partial \epsilon} \frac{\partial \epsilon}{\partial T}$ containing the differentials with respect to both Boltzmann-factor and dielectric terms. In eq. (24), the first term on the right-hand side provides the heat-capacity change associated with the rearrangement of solvent ions, while the second term provides the heat-capacity change associated with aqueous dipole reorientation, and the cross terms provide the coupling between these two contributions.

Using this approach, the contribution of electrostatics to the ΔC_p associated with binding for DNA binding reactions involving the ligands DAPI, netropsin, lexitropsin, and the λ repressor has been studied (Gallagher and Sharp, 1997). In general, the heat-capacity change due to electrostatics is small. Overall the electrostatic interactions contribute a positive heat-capacity change in binding in the range 15–90 cal/mole/K. This is dominated by a positive term arising from aqueous dipole rearrangement, and it is opposed by the contribution of mobile solvent ions and the coupling term.

2.6. Association Rates

Association rates can be greatly affected by the long-range electrostatic forces acting between the protein and the ligand or substrate. The general approach to calculating the bimolecular first-encounter association rate constant for the reaction A+B→B is based on the formula (Allison et al., 1985; Northrup et al., 1984)

$$k_a = 4\pi D_{ab} R_{init} P_{ab}, \qquad (25)$$

where D_{ab} is the mutual translational diffusion constant of A and B ($D_{ab} = D_a + D_b$), R_{init} is some initial encounter distance between A and B that is large enough for the electrostatic forces between A and B to be negligible, while P_{ab} is the probability that the substrate, having approached to distance R_{init}, will collide productively with the protein, which can be estimated through Brownian-dynamic (BD) simulations. Substrate trajectories are generated using a combination of a diffusive random walk (characterized by the rotational and translational diffusion constants) combined with orientational and translational drift produced by, for example, electrostatic forces (Ermak and McCammon, 1978). A trajectory is propagated until either a productive collision occurs or the substrate–enzyme separation exceeds some cutoff $R_{cut} > R_{init}$. Many such trajectories are run until an estimate (corrected for the probability that the substrate will reenter after the cutoff (Allison et al., 1985; Northrup et al., 1984)) of the fraction of productive trajectories, P_{ab}, is obtained to the required precision. Electrostatic body forces are usually calculated using the test-charge approach (eq. (8)) from pre-computed electrostatic potential distributions obtained from the PB equation (Sharp et al. 1987a,b). This is rapid, which is necessary given the many hundreds of thousands of force evaluations required by BD simulations. It neglects, however, the effect that one molecule may have on the potential from the other (Davis et al., 1991; Davis and McCammon, 1990). Although current computer resources are insufficient to implement the exact force calculation (e.g. using eq. (7)), many useful results can be obtained from such BD simulations. Sophisticated simulations can incorporate adaptive step sizes, rotational diffusion (Andrew et al., 1993; Northrup et al., 1987), hydrodynamic effects (Allison et al., 1986), and the flexibility of substrate or enzyme (Luty et al., 1993; Wade et al., 1993). The combined PB/BD approach has been used to study the effects of ionic strength and chemical modification on SOD turnover (Davis et al., 1991; Sharp et al., 1987a), to predict the effect of charged residue modifications (Getzoff et al., 1992; Sharp et al., 1987a), to study orientational and ionic-strength effects in interactions between cytochrome-c and cytochrome-c peroxidase (Andrew et al., 1993; Northrup et al., 1993), to study antibody–antigen interactions (Holst et al., 1994b), to look at interactions between acetyl choline and acetyl choline esterase (Antosiewicz et al., 1995), and to look at diffusion of substrates between bifunctional enzyme active sites (Elcock et al., 1996).

3. Summary

Electrostatic interactions help determine and maintain the structures of proteins and nucleic acids. These interactions affect the reactivity and functional properties of proteins and nucleic acids when movement of charged or dipolar groups and molecules are involved. The relationship between the electrostatic potential and various macromolecular properties of current interest was described using a combination of classical electrostatics, thermodynamics, and statistical mechanics. The Poisson–Boltzmann model for electrostatics was chosen as the theoretical framework to illustrate how one may calculate these electrostatic interactions.

References

Allison, S. A., Northrup, S. H., and McCammon, J. A. (1985) *J. Chem. Phys.* **83**, 2894.

Allison, S. A., Northrup, S. H., and McCammon, J. A. (1986) *Biophysical J.* **49**, 167.

Andrew, S., Thomasson, K., and Northrup, S. (1993) *Journal of the American Chemical Society* **115**, 5516–21.

Antosiewicz, J., Gilson, M. K., Lee, I. H., and McCammon, J. A. (1995) *Biophys. J.* **68**, 62–8.

Åqvist, J., Luecke, H., Quiocho, F. A., and Warshel, A. (1991) *Proc. Natl. Acad. Sci. USA* **88**, 2026–30.

Bacquet, R., and Rossky, P. (1984) *J. Phys. Chem.* **88**, 2660.

Bashford, D., and Gerwert, K. (1992) *J. Mol. Biol.* **224**, 473–86.

Bashford, A., and Karplus, M. (1990) *Biochemistry* **29**, 10219–25.

Bashford, D., Karplus, M., and Canters, G. W. (1988) *J. Mol. Biol.* **203**, 507.

Beroza, P., Fredkin, D., Okamura, M., and Feher, G. (1991) *Proc. Natl. Acad. Sci. USA* **88** 5804–8.

Bharadwaj, R., Windemuth, A., Sridharan, S., Honig, B., and Nicholls, A. (1994) *J. Comp. Chem.* **16**, 898–913.

Davis, M. E. (1994) *J. Chem. Phys.* **100**, 5149–59.

Davis, M. E., Madura, J. D., Sines, J., Luty, B. A., Allison, S. A., and McCammon, J. A. (1991) *Meth. Euz.* **202**, 473–97.
Davis, M. E., and McCammon, J. A. (1989) *J. Comp. Chem.* **10**, 386–95.
Davis, M. E., and McCammon, J. A. (1990) *Journal of Computational Chemistry* **11**, 401–9.
Elcock. A. H., Potter, M. J., Matthews, D. A., Knighton, D. R., and McCammon, J. A. (1996) *J. Mol. Biol.* **262**, 370–4.
Ermak, D. L., and McCammon, J. A. (1978) *J. Chem. Phys.* **69**, 1532.
Ewing, P. J., and Lybrand, T. P. (1993) *J. Phys. Chem.* **98**, 1748–52.
Gallagher, K., and Sharp, K. A. (1997) *Biophys. J.* **A72**, 95.
Getzoff, E., Cabelli, D., Fisher, C., Parge, H., Viezzoli, M., Banci, L., and Hallewell, R. (1992) *Nature* **358**, 347–51.
Gilson, M. (1993) *Proteins—Structure Function and Genetics* **15**, 266–82.
Gilson, M., Davis, M., Luty, B., and McCammon, J. (1993) *J. Phys. Chem.* **97**, 3591–600.
Gilson, M. and Honig, B. (1986) *Biopolymers* **25**, 2097–119.
Gilson, M., and Honig, B. (1987) *Nature* **330**, 84.
Gilson, M., McCammon, J., and Madura, J. (1995) *J. Comp. Chem.* **16**, 1081–95.
Gilson, M., Sharp, K. A., and Honig, B. (1988) *J. Comp. Chem.* **9**, 327–35.
Gunner, M., and Honig, B. (1990) In: *Perspectives in Photosynthesis*, J. Jortner and B. Pullman (eds.). Dordrecht: Kluwer Academic Publishers. Pp. 53–60.
Gunner, M. R., and Honig, B. (1991) *Proc. Natl. Acad. Sci. USA* **88**, 9151–5.
Holst, M., Kozack, R., Saied, F., and Subramaniam, S. (1994a) *J. Biomol. Struct. Dynam.* **11**, 1437–45.
Holst, M., Kozack, R. E., Saied, F., and Subramaniam, S. (1994b) *Proteins* **18**, 231–45.
Holst, M., and Saied, F. (1993) *J. Comp. Chem.* **14**, 105–13.
Jayaram, B., Fine, R., Sharp, K. A., and Honig, B. (1989a) *J. Phys. Chem.* **93**, 4320–7.
Jayaram, B., Sharp, K. A., and Honig, B. (1989b) *Biopolymers* **28**, 975–93.
Jean-Charles, A., Nicholls, A., Sharp, K., Honig, B., Tempczyk, A., Hendrickson, T., and Still, C. (1990) *J. Am. Chem. Soc.* **113**, 1454–5.
Langen, R., Jensen, G., Jacob, U., Stephens, P., and Warshel, A. (1992) *Journal of Biological Chemistry* **267**, 25625–7.
Langsetmo, K., Fuchs, J. A., Woodward, C., and Sharp, K. A., (1991) *Biochemistry* **30**, 7609–14.
Lee, F. S., Chu, Z. T., Bolger, M. B., and Warshel, A. (1992) *Protein Engineering* **5**, 215–28.
Lide, D. R. (1990). *CRC Handbook of Chemistry and Physics.*
Luty, B., Wade, R., Madura, J., Davis, M., Briggs, J., and McCammon, J. (1993) *Journal of Physical Chemistry* **97**, 233–7.
Marcus, R. (1956) *J. Chem. Phys.* **24**, 966–78.

Misra, V., Hecht, J., Sharp, K., Friedman, R., and Honig, B. (1994a) *J. Mol. Biol.* **238**, 264–80.
Misra, V., and Honig, B. (1995) *Proc. Natl. Acad. Sci. USA* **92**, 4691–5.
Misra, V., Sharp, K., Friedman, R., and Honig, B. (1994b) *J. Mol. Biol.* **238**, 245–63.
Mohan, V., Davis, M. E., McCammon, J. A., and Pettitt, B. M. (1992) *Journal of Physical Chemistry* **96**, 6428–31.
Murthy, C. S., Bacquet, R. J., and Rossky, P. J. (1985) *J. Phys. Chem.* **89**, 701.
Nakamura, H., Sakamoto, T., and Wada, A. (1988) *Protein Engineering* **2**, 177–83.
Nicholls, A., and Honig, B. (1991) *J. Comp. Chem.* **12**, 435–45.
Northrup, S., Thomasson, K., Miller, C., Barker, P., Eltiss, L., Guillemette, J., Inglis, S., and Mauk, A. (1993) *Biochemistry* **32**, 6613–23.
Northrup, S. H., Allison, S. A., and McCammon, J. A. (1984) *J. Chem. Phys.* **80**, 1517.
Northrup, S. H., Boles, J. O., and Reynolds, J. C. (1987) *J. Phys. Chem.* **91**, 5991.
Oberoi, H., and Allewell, N. (1993) *Biophysical Journal* **65**, 48–55.
Olmsted, M. C., Anderson, C. F., and Record, M. T. (1989) *Proc. Natl. Acad. Sci. USA* **86**, 7766–70.
Olmsted, M. C., Anderson, C. F., and Record M. T. (1991) *Biopolymers* **31**, 1593–604.
Pack, G., Garrett, G., Wong, L., and Lamm, G. (1993) *Biophys. J.* **65**, 1363–70.
Pack, G. R., and Klein, B. J. (1984) *Biopolymers* **23**, 2801.
Pack, G. R., Wong, L., and Prasad, C. V. (1986) *Nucl. Acid Res.* **14**, 1479.
Rashin, A., and Bukatin M., (1993) *Journal of Physical Chemistry* **97**, 1974–9.
Rashin, A. A. (1990) *J. Phys. Chem.* **94**, 725–33.
Rashin, A. A., and Honig, B. (1985) *J. Phys. Chem.* **89**, 5588–93.
Rashin, A. A., and Namboodiri, K. (1987) *J. Phys. Chem.* **91**, 6003.
Record, T., Olmsted, M., and Anderson, C. (1990) *Theoretical Biochemistry and Molecular Biophysics*. Adenine Press.
Reiner, E. S., and Radke, C. J. (1990) *J. Chem. Soc. Faraday Trans.* **86**, 3901.
Sampogna, R., and Honig, B. (1994) *Biophysical J.* **66**, 1341–52.
Schaeffer, M., and Frommel, D. (1990) *J. Mol. Biol.* **216**, 1045–66.
Sharp, K., Fine, R., and Honig, B. (1987a) *Science* **236**, 1460.
Sharp, K., Fine, R., Schulten, K., and Honig, B. (1987b) *J. Phys. Chem.* **91**, 3624.
Sharp, K., and Honig, B. (1990) *J. Phys. Chem.* **94**, 7684–92.
Sharp, K. A. (1995) *Biopolymers* **36**, 227–43.
Sharp, K. A. (1998) *Biophys. J.* **73**, 1241–50.

Sharp, K. A., Friedman, R., Misra, V., Hecht, J., and Honig, B. (1995) *Biopolymers* **36**, 245–62.

Sharp, K. A., Jean-Charles, A., and Honig, B. (1992) *J. Phys. Chem.* **96**, 3822–8.

Simonson, T., and Brunger, A. (1994) *J. Phys. Chem.* **98**, 4683–94.

Simonson, T., and Perahia, D. (1995) *Proc. Natl. Acad. Sci. USA* **92**, 1082–6.

Sitkoff, D., Sharp, K., and Honig, B. (1994) *J. Phys. Chem.* **98**, 1978–88.

Slagle, S., Kozack, R. E., and Subramaniam, S. (1994) *J. Biomol. Struct. Dyn.* **12**, 439–56.

Smith, P., Brunne, R., Mark, A., and vanGunsteren, W. (1993) *Journal of Physical Chemistry* **97**, 2009–14.

Sternberg, M., Hayes, F., Russell, A., Thomas, P., and Fersht, A. (1987) *Nature* **330**, 86.

Still, C., Tempczyk, A., Hawley, R., and Hendrickson, T. (1990) *J. Am. Chem. Soc.* **112**, 6127–9.

Takashima, S., and Schwan, H. P. (1965) *J. Phys. Chem.* **69**, 4176.

Wade, R., Davis, M., Luty, B., Madura, J., and McCammon, J. (1993) *Biophysical Journal* **64**, 9–15.

Warwicker, J. (1994) *J. Mol. Biol.* **236**, 887–903.

Warwicker, J., and Watson, H. C. (1982) *J. Mol. Biol.* **157**, 671–9.

Wendoloski, J. J., and Matthew, J. B. (1989) *Proteins* **5**, 313.

Yang, A., Gunner, M., Sampogna, R., Sharp, K., and Honig, B. (1993) *Proteins—Structure Function and Genetics* **15**, 252–65.

Yang, A., and Honig, B. (1994) *J. Mol. Biol.* **237**, 602–14.

Yoon, L., and Lenhoff, A. (1992) *J. Phys. Chem.* **96**, 3130–4.

Zauhar, R., and Morgan, R. J. (1985) *J. Mol. Biol.* **186**, 815–20.

Zhou, H. X. (1994) *J. Phys. Chem.* **100**, 3152–62.

5

Thermodynamics of Formation of Secondary Structure in Nucleic Acids

Ignacio Tinoco, Jr. and Michael Schmitz

We want to understand the thermodynamic parameters that characterize the conformations of nucleic acids. In biological cells, RNA and DNA molecules are present as compact folded structures dependent on Watson–Crick base pairing and other molecular interactions. The DNA in the nucleus is mainly in the form of chromosome-sized (approximately 100 megabase pairs) double-stranded B-form helices. However, portions of the duplex are continually melting and reforming as the DNA is replicated, and is transcribed into RNA. Genomic DNA is associated with proteins in the nucleus, and duplex melting at physiological temperatures is facilitated by helicase enzymes. The kinetics of all the interactions are of obvious importance. Nonetheless, the thermodynamic stability of a duplex region relative to melted single strands depends on the base sequence of the region. The RNA that is being synthesized folds into Watson–Crick paired structures, but the pairing is interrupted by mismatches, loops, and bulges. These specific structures are recognized by the many proteins used in processing the RNA transcripts into messenger, ribosomal, and transfer RNAs, and other molecules, and in transporting the RNAs to the sites of their biological activities. We recognize that the interactions of the RNAs with proteins are essential to their function. However, the RNAs also interact with other RNAs, such as ribozymes, and interaction among nucleic acids is crucial in antisense mechanisms. We will concentrate on the intermolecular and intramolecular interactions of the nucleic acids—both DNA and RNA. We will review what is known about the thermodynamics of forming base-paired structures in nucleic acids, and how the data have been obtained and interpreted.

1. Fundamentals: Definitions, Activities, Standard States

We start by defining the general thermodynamic equations we will use, and by introducing our notation. For any reaction the standard free-energy change $\Delta G^0(T)$, the standard enthalpy change $\Delta H^0(T)$, and the standard entropy change $\Delta S^0(T)$ are related at any temperature T by

$$\Delta G^0(T) = \Delta H^0(T) - T\Delta S^0(T). \tag{1}$$

The changes in thermodynamic variables refer to reactants in their standard states going to products in their standard states. The standard state is the solute standard states of 1 M concentration of species (extrapolated from very dilute solution) at 1 atm pressure. The standard free-energy change can be obtained from the equilibrium constant $K(T)$ for the reaction, and the standard enthalpy change can be obtained from its temperature derivative:

$$\Delta G^0(T) = -RT \ln K(T), \tag{2}$$

$$\Delta H^0(T) = -R\frac{\partial \ln K(T)}{\partial (1/T)} = RT^2 \frac{\partial \ln K(T)}{\partial T}. \tag{3}$$

Equation (3) is called the van't Hoff equation. The standard entropy change, $\Delta S^0(T)$ is obtained from eq. (1).

For the simple reaction

$$A \rightarrow B,$$

the equilibrium constant K is the ratio of the equilibrium activities at a chosen temperature T and at a pressure of 1 atm:

$$K(T) = \frac{a_B}{a_A}. \tag{4}$$

Activities are defined relative to a standard state—where the activity is equal to 1. For equilibria involving nucleic acids, it is common to specify a solvent composition—for example, 1 M NaCl with 10 mM pH 7 phosphate buffer—as part of the standard state. The standard state for the nucleic acid species is 1 M concentration (extrapolated from infinite dilution) in the chosen solvent, the chosen temperature, and 1 atm pressure. We are thus defining the activity as

$$a = \gamma c, \tag{5}$$

where the activity coefficient γ is equal to 1 in the limit as concentration c approaches 0 (the limit of infinite dilution). The standard state has unit activity at 1 M concentration, but with the properties of infinite dilution ($\gamma = 1$).

The other solutes are considered part of the solvent, although some of them may be involved in the equilibrium. For example, H^+, Mg^{2+}, and buffer species can all have differential binding to products and reactants, and are sometimes explicitly included in the reaction. It is clearly important to specify the exact equilibrium under consideration when a value for K or ΔG^0 is reported.

To measure the equilibrium constant experimentally, one assumes that the nucleic-acid concentration is dilute enough for concentrations to replace activities. The equilibrium constant is then the ratio of equilibrium concentrations:

$$K(T) = \frac{c_B}{c_A}. \tag{6}$$

One should verify that the equilibrium constant is indeed independent of nucleic-acid concentration by making measurements at different (dilute) nucleic-acid concentrations. The equilibrium constant should depend only on temperature and solvent composition. One can thus obtain the desired thermodynamic variables of a reaction by measuring the equilibrium concentrations of the species over a range of temperatures in the choice of solvent.

The key requirement for obtaining accurate values of $\Delta G^0(T)$, $\Delta H^0(T)$ and $\Delta S^0(T)$ for a reaction from equilibrium constants is the ability to measure the equilibrium concentrations of the species involved in the reaction. An analytical method that is specific for each species in the reaction should be used. When less specific methods, such as ultraviolet absorbance, are used, it is important to test assumptions about the equilibria involved. Note that $K(T)$ and ΔG^0 are the primary quantities calculated, so ΔG^0 has the highest precision among ΔG^0, ΔH^0, and ΔS^0. The enthalpy depends on the temperature dependence of $K(T)$, and the entropy depends both on $K(T)$ and on its temperature dependence.

2. Fundamentals: Calorimetry

Thermodynamics of reactions can be measured in a calorimeter by isothermal calorimetry or by differential scanning calorimetry. In isothermal calorimetry, the heat evolved or absorbed on mixing solutions of reactants is measured at constant temperature. In differential scanning calorimetry, the heat absorbed is measured as the temperature is scanned over a temperature range in which the reaction occurs.

Isothermal Calorimetry

The enthalpy change for any process is equal to the heat evolved or absorbed at constant pressure:

$$\Delta H = q_P. \tag{7}$$

For a chemical reaction, the reactants can be mixed in a calorimeter, and the heat effect measured from the *small* temperature change that occurs in the sample in the mixing chamber:

$$q_P = C_P(\text{calorimeter})\Delta T.$$

The heat capacity at constant pressure of the sample and chamber, C_P (calorimeter), is determined by adding calibrated amounts of electrical energy. Therefore, from the temperature change ΔT of the sample, q_P is obtained. The heat effect, and thus the enthalpy change, corresponds to the reaction that actually occurs in the calorimeter at a chosen, essentially constant, temperature. The reactants at the initial concentrations react to form the equilibrium concentrations of products and rectants in the final solution. If the reaction is essentially complete—the equilibrium constant is very large—the measured ΔH is the heat of the reaction for the number of moles of reactants in the chamber. Of course, it is important to know what reaction actually occurs in the calorimeter. Analysis of the contents of the calorimeter is necessary to determine the stoichiometry of the reaction.

The heat effects should be measured with concentrations of nucleic acids dilute enough for a standard ΔH^0 to be obtained. The standard ΔH^0 per mole of reaction is designated as $\Delta \overline{H}^0$. Presumably the heats of dilution on mixing the reactants are negligible in relation to the heat of reaction, and the solvents for the reactants are identical, so that there is no heat effect of solvent mixing.

An equilibrium constant for the reaction can be obtained by titration; one reactant is added in small increments to the other reactant. The heat evolved or absorbed after each increment of reactant is added is directly proportional to the extent of reaction. Consider the bimolecular reaction

$$A + B \rightarrow C.$$

The molar enthalpy of the reaction can be written in terms of the molar enthalpies of products and reactants:

$$\Delta \overline{H} = \overline{H}_C - \overline{H}_B - \overline{H}_A. \tag{8}$$

In the microcalorimeter cell a heat release of q means that $n = q/\Delta \overline{H}$ mols of reaction occurred; a concentration $x = n/V_S$ of product C formed, where V_S is the volume of solution in the cell. The equilibrium constant for the reaction is

$$K = \frac{[C]}{[A][B]} = \frac{x}{(A_o - x)(B_o - x)}, \tag{9}$$

where A_o and B_o are the concentrations of reactants added to the calorimeter. The sum of the heat q released after each addition of reactant provides x and thus the equilibrium constant.

Isothermal calorimetry can provide all the thermodynamic parameters: ΔH^0 from the total heat evolved, $\Delta G^0 = -RT \ln K$ from the increments of heat evolved during the titration of one reactant with another, and ΔS^0 from $(\Delta H^0 - \Delta G^0)/T$. The isothermal calorimetry experiments can be done at a series of temperatures to obtain the temperature dependences of the thermodynamic parameters.

Differential Scanning Calorimetry

The enthalpy change for a thermal transition can be measured by integrating the heat effect at constant pressure between two temperatures that span the transition. To measure the enthalpy change of a chemical reaction, the solvent is placed in one chamber and an equal amount of solution is placed in an adjoining chamber. As the temperature is raised, the reaction progresses, which means that the reaction is endothermic—heat is absorbed. For example, a double strand melts to single strands. The excess heat necessary to raise the temperature of the solution—where the reaction occurs—is measured. This is differential scanning calorimetry. In mathematical terms,

$$\Delta H_{T_1 \to T_2} = \int_{T_1}^{T_2} dq_\mathrm{P}. \tag{10}$$

The enthalpy of the thermal transition indicated in eq. (1) is the excess enthalpy change for the solution relative to pure solvent. Consider a reaction A → B that has a negative ΔH, so that raising the temperature favors A (the equilibrium constant decreases with T). An example could be the folding of a single strand into a hairpin. A plot of the extent of reaction as a function of temperature—the fraction f_A of molecules that are A, and the fracation f_B that are B—is shown in fig. 5.1. The melting temperature T_m is defined as the temperature for which $f_\mathrm{A} = f_\mathrm{B} = 0.5$. If T_1 and T_2 cover the range in which the reaction is complete, then the integration in eq. (10) is approximately equal to

$$\int_{T_1}^{T_2} dq_\mathrm{P} = \Delta H_{\mathrm{A} \to \mathrm{B}}(T_\mathrm{m}) + C_\mathrm{PB}(T_\mathrm{m} - T_1) + C_\mathrm{PA}(T_2 - T_\mathrm{m}), \tag{11}$$

where $\Delta H_{\mathrm{A} \to \mathrm{B}}(T_\mathrm{m})$ is the enthalpy change of reaction at T_m, and C_PA and C_PB are the heat capacities at constant pressure of A and B, assumed independent of T, respectively. The total heat absorbed is the sum of the enthalpy change of the reaction at T_m plus the heat of raising the temperature of B from T_1 to T_m and of A from T_m to T_2. Equation (11) would be exact if the reaction occurred as a step function at T_m. However, corrections for the actual shape of f against T (fig. 5.1) are usually negligible.

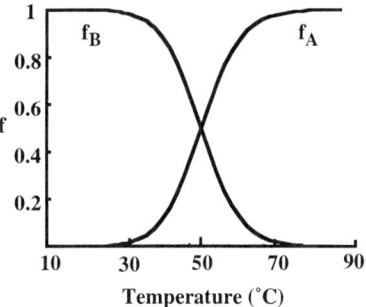

Figure 5.1. For reaction A→B, the fractions f_A and f_B of molecules that are A and B, respectively, are plotted against temperature (°C). The melting temperature $T_m = 323$ K $= 50$°C is defined as the temperature for which $f_A = f_B = \frac{1}{2}$. The curves were calculated from eqs. (15, 17) with values of $\Delta H^0 = -20{,}000\,R$ (J) and $\Delta S^0 = -20{,}000R/323$ (J K^{-1}).

We can derive eq. (11) and analyze the heat effects more closely by considering the differential change in enthalpy for a differential change in temperature. The differential form of eq. (10) is written as

$$dH = \left[\frac{\partial H}{\partial T}\right]_P dT = C_P\, dT, \tag{12}$$

where C_P is the excess heat capacity at constant pressure of solution relative to solvent. The excess heat capacity depends explicitly on the heat of reaction and on the heat capacities of reactants and products. It multiplies the temperature differential in the equation

$$dH = C_P\, dT = \left(\Delta H_{A \to B}(T)\frac{df_B}{dT} + C_{PA}f_A(T) + C_{PB}f_B(T)\right) dT. \tag{13}$$

The temperature dependence of $\Delta H_{A \to B}(T)$ is

$$\Delta H_{A \to B}(T) = \Delta H_{A \to B}(T_m) + (C_{PB} - C_{PA})(T - T_m). \tag{14}$$

The temperature dependence of the excess heat capacity depends on the temperature dependence of the equilibrium constant:

$$f_A(T) = \frac{1}{1 + K(T)}, \qquad f_B(T) = 1 - f_A(T) = \frac{K(T)}{1 + K(T)}, \tag{15}$$

$$\frac{df_A}{dT} = -\frac{df_B}{dT} = \frac{-K(T)}{[1 + K(T)]^2}\frac{\Delta H^0(T)}{RT^2}. \tag{16}$$

Integrating eqs. (13–16) between two temperatures that encompass the range in which f_A and f_B vary from 0 to 1 yields the equivalent of eq. (11). To illustrate the shape of an excess heat capacity curve, one can use temperature-independent values of ΔH^0 and ΔS^0 to write $K(T)$ as

FORMATION OF SECONDARY STRUCTURE IN NUCLEIC ACIDS

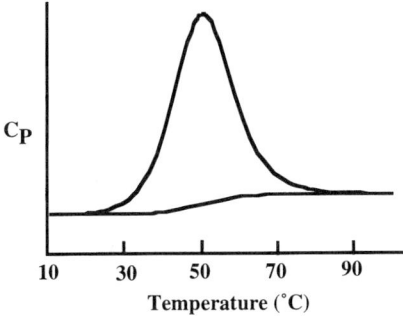

Figure 5.2. The excess heat capacity C_P (heat capacity of solution minus heat capacity of solvent) is plotted against T for the reaction of A → B shown in Fig. 5.1. Equation (13) is used with $\Delta H^0 = -20{,}000R$ (J), $\Delta S^0 = -20{,}000R/323$ (J K^{-1}), $C_{PA} = 300R$ (J K^{-1}), and $C_{PB} = 200R$ (J K^{-1}). The baseline corresponds to heating reactants and products.

$$K(T) = e^{\Delta S^0/R} e^{-\Delta H^0/RT}. \qquad (17)$$

Choosing values of $\Delta H^0 = -20{,}000R$ (J) and $\Delta S^0 = -20{,}000R/323$ (J K^{-1}) places the T_m at 323 K (50°C) for the A → B equilibrium. To illustrate the different contributions to the measured excess heat capacity, the solute heat capacities are chosen as $C_{PA} = 300R$ and $C_{PB} = 200R$. The shape of the excess heat capacity against temperature is shown in fig. 5.2. The baseline shows the heat absorbed required to heat the reactants and products; the amount of heat depends on their heat capacities. Between T_1 and the melting temperature, mainly B is being heated; from T_m to T_2, mainly A is being heated. The contribution of the enthalpy of the reaction is a maximum at T_m. The enthalpy change of the reaction is the integral of the heat evolved after subtracting the baseline, as indicated by eq. (11). It is the area between the curves in fig. 2. Experimentally, one usually draws the baseline by linear extrapolation between estimated end points for the reaction.

The shape of the curve for excess heat capacity, shown in fig. 5.2, was calculated with temperature-independent values of ΔH^0 and ΔS^0. Using temperature-dependent expressions for ΔH^0 and ΔS^0 consistent with the different heat capacities for A and B would not appreciably change the shape.

The entropy change for a thermal transition can be measured from the integral of the *reversible* heat effect at each temperature divided by that temperature. The integration is between two temperatures that span the transition.

$$\Delta S_{T_1 \to T_2} = \int_{T_1}^{T_2} \frac{dq_{\text{rev P}}}{T} = \int_{T_1}^{T_2} \frac{C_P}{T}\, dT. \qquad (18)$$

The integral is the sum of the entropy change of the reaction at T_m and the entropy changes in heating reactants and products:

$$\int_{T_1}^{T_2} \frac{C_P}{T} dT = \Delta S_{A\to B}(T_m) + C_{PB} \ln \frac{T_m}{T_1} + C_{PA} \ln \frac{T_2}{T_m}, \quad (19)$$

where $\Delta S_{A\to B}(T_m)$ is the entropy of reaction at T_m. Experimentally, the entropy is obtained by dividing the measured C_P by the temperature, then integrating the area of C_P/T with respect to T, similarly to what is done in fig. 2. The entropy at any temperature is given by

$$\Delta S_{A\to B}(T) = \Delta S_{A\to B}(T_m) + (C_{PB} - C_{PA}) \ln \frac{T}{T_m}. \quad (20)$$

The free energy $\Delta G(T)$ of the reaction at any T is obtained from eq. (1).

It is straightforward in principle to measure heat effects on mixing two solutions, or for raising the temperature of a sample. However, to obtain accurate values of ΔH^0, ΔS^0, and ΔG^0 for a reaction requires that we know what reaction occurs. If we mix solutions of reactants, we need to know what the products are, and what quantities of the reactants (if any) are left. In a differential scanning calorimeter (DSC) experiment, we need to know what species are present initially at T_1, and what species are present finally at T_2. To interpret the measured heat effects, we need to understand the equilibria involved. Note that, in calorimetry, because heat is measured directly, ΔH^0 has the highest precision among ΔH^0, ΔS^0, and ΔG^0. In a mixing experiment, ΔH^0 can be measured at different temperatures. In a scanning experiment, ΔH^0 is obtained at the T_m; but, when C_{PA} and C_{PB} are measured, ΔH^0 can be calculated at any T from eq. (14). The entropy change ΔS^0 can be obtained from a scanning experiment, but the measurement is less direct than that for ΔH^0.

3. Experimental Methods: Absorption

All-or-none (Two-state) Equilibria

The fundamental thermodynamic equations given above are mostly identities, true by definition. However, the application of the equations to experimental data is not always obvious. The macroscopic thermodynamic equations need to be applied to molecular species. In principle, activities can be measured by several methods, but for nucleic acids one is usually constrained to measure concentrations.

Concentrations are usually measured from absorbance data. Nucleic acids have strong absorption bands near 260 nm, and the absorption increases when stacked bases in double-stranded helices become unstacked (Gray et al., 1995). The increase in absorbance as temperature is increased is an absorption melting curve. It can provide standard thermodynamic parameters. The optimum wavelength to choose for a melting curve is the one that gives the largest change on melting; it depends on the base composition of the nucleic acid. Molecules high in G–C content are best studied at 280 nm; those high in A–T or A–U at 260 nm (Felsenfeld and Hirschman, 1965). The relative amounts of G–C and A–U or A–T base pairs melting in any transition can be obtained from melting curves recorded at both wavelengths. This additional information can be helpful especially in the analysis of more complex melting profiles.

Consider the general equilibrium reaction of n strands of a nucleic acid forming an n-stranded structure.

$$nS \to P: \quad K = \frac{[P]}{[S]^n}. \qquad (21)$$

Mass balance gives

$$[S] + n[P] = c_t,$$

with c_t equal to the total strand concentration in mol ℓ^{-1}. Defining f as the fraction of the strands in the n-stranded species, one obtains

$$K = \frac{f}{n\, c_t^{n-1}(1-f)^n} \qquad (22)$$

with

$$f = \frac{n[P]}{c_t} \quad \text{and} \quad 1-f = \frac{[S]}{c_t}. \qquad (23)$$

The equilibrium reaction in eq. (21) can represent a single strand folding into a hairpin ($n = 1$), a single strand converting to a double-stranded helix ($n = 2$), or a single strand converting to a triplex, quadruplex, or n-plex form. However, it always defines an *all-or-none transition*—a two-state transition. There are no intermediate species in the equilibrium; all the species can be identified as either single-stranded or n-stranded. The concentrations of the two species, and thus the equilibrium constant, can be calculated from the single variable f. The absorbance of the equilibrium mixture is written using the Beer–Lambert equation $A = \varepsilon c l$, with ε the extinction coefficient, c the concentration, and l the path length. In a 1-cm cell:

$$A(T) = \varepsilon_S[S] + \varepsilon_P[P], \qquad (24)$$

where ε_S and ε_P, are extinction coefficients per mole of single-stranded and n-stranded species, respectively. Using eq. (23) to replace concentrations by f, one obtains

$$A(T) = (1-f)A_S(T) + fA_P(T) \qquad (25)$$

$$A_S(T) + \varepsilon_S(T)c_t \quad \text{and} \quad A_P(T) = \varepsilon_P(T)\frac{c_t}{n} \qquad (26)$$

with $A_S(T)$ and $A_P(T)$ the temperature-dependent absorbances of the two species.

Thus, from an absorbance melting curve, the fraction f of strands in the n-stranded structure is

$$f = \frac{A(T) - A_S(T)}{A_P(T) - A_S(T)}. \tag{27}$$

For a two-state (i.e. all-or-none) equilibrium, the concentration of the two species, given by eq. (23), and the equilibrium constant at any temperature, given by eq. (22), can be obtained from the absorbance melting curve of $A(T)$ against T, using eq. (27).

An absorbance melting curve for a hairpin is shown in fig. 5.3, constructed from the same thermodynamic parameters as used in figs. 5.1–5.2. The temperature–dependent absorbances of single strand and hairpin are assumed linear, with a 20% difference in absorbance at 0°C, and a higher slope for the absorbance of the single strands. Extrapolation of upper and lower baselines leads to f from eq. (27) and $K(T)$ from eq. (22). The standard thermodynamic parameters of $\Delta G^0(T)$, $\Delta H^0(T)$, and $\Delta S^0(T)$ are then calculated. One can instead directly fit the absorbance melting curve to linear upper and lower baselines and temperature-independent values of ΔH^0 and ΔS^0. A nonlinear least-squares fit of the melting curve to six parameters (ΔH^0, ΔS^0, and the slopes and intercepts for the upper and lower baselines) is used (Petersheim and Turner, 1983).

The Slope of f *with Respect to Temperature*

There are many ways to obtain thermodynamic parameters from a melting curve; the key quantity is f—the fraction of strands in a structured state—obtained from eq. (27). The equilibrium constant. $K(T)$ and ΔG^0 can be obtained from eq. (22) and eq. (2). The enthalpy change ΔH^0 can be obtained from the temperature derivative of $K(T)$ at constant total concentration of strands, c_t. The slope $(\partial f/\partial T)$ of f against T obtained from a melting curve at a constant value of total strands gives ΔH^0.

$$\Delta H^0(T) = RT^2 \frac{1 + (n-1)f}{f(1-f)} \frac{\partial f}{\partial T}. \tag{28}$$

The value of ΔH^0 can be calculated from eq. (28) at any temperature for which precise values of f and $\partial f/\partial T$ can be obtained. Usually ΔH^0 is calculated at T_m where $f = \frac{1}{2}$; then

$$\Delta H^0(T_m) = 2RT_m^2(n+1)\frac{\partial f}{\partial T}(T_m). \tag{29}$$

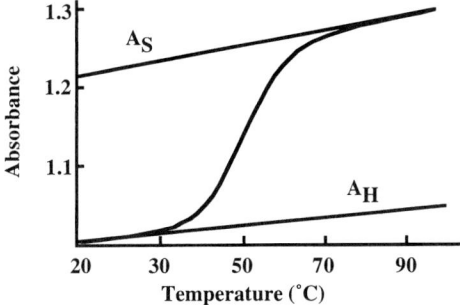

Figure 5.3. An absorbance melting curve for a two-state transition. We consider the equilibrium of a single strand to hairpin (S → H) with $\Delta H^0 = -20{,}000R$ (J) and $\Delta S^0 = -20{,}000R/323$ (J K^{-1}). The temperature dependences of the absorbance of the single strands and of the hairpins are: $A_S(T) = 1.20 + 0.001(T - 273)$, and $A_H(T) = 1.00 + 0.0005(T - 273)$.

We note that the maximum value $(\partial f/\partial T)_{max}$ of $\partial f/\partial T$ occurs at the melting temperature, when $f = \frac{1}{2}$, only for $n = 1$—a transition of a single strand to a hairpin, for example. In general, the maximum value of $\partial f/\partial T$ occurs for a value of f less than $\frac{1}{2}$. The formula for $n > 1$ is

$$f \text{ at } \left(\frac{\partial f}{\partial T}\right)_{max} = \frac{1 - \sqrt{n}}{1 - n}. \tag{30}$$

Thus for the transition of a single strand to a double strand (i.e., $n = 2$), the maximum slope of f against T occurs at $f = 0.4142$.

A simple approximation can be made to obtain ΔH^0 for a unimolecular melting, such as a hairpin to a single strand. For a sharp transition, $\partial f/\partial T$ is directly proportional to $-\partial A/\partial T$, because the effect of base lines is negligible. The negative sign follows from the increase in absorbance as the fraction of helix decreases. If the shape of the $\partial A/\partial T$ curve is approximated as a symmetric triangle, eq. (29) is approximated by

$$\Delta H^0(T_m) = -4RT_m^2/\Delta T_{1/2}, \tag{31}$$

where $\Delta T_{1/2}$ is the full width of the $\partial A/\partial T$ curve at half-height. This equation can be used to estimate $\Delta H^0(T_m)$ directly from $\partial A(T)/\partial T$ without first calculating the fraction f of base-paired strands (Riesner et al., 1973).

The Plot of ln c_t Against $1/T_m$

The enthalpy of reaction can be obtained from the concentration dependence of the temperature required to form a certain fraction of product. Solving for T in the equation relating $K(T)$ to ΔH^0 and ΔS^0, we have

$$\frac{1}{T} = \frac{\Delta S^0}{\Delta H^0} - \frac{R}{\Delta H^0} \ln K(T), \tag{32}$$

and using eq. (22) for K, one obtains

$$\frac{1}{T} = \frac{R}{\Delta H^0}(n-1)\ln c_t + \frac{\Delta S^0}{\Delta H^0} - \frac{R}{\Delta H^0} \ln \frac{f}{n(1-f)^n}. \tag{33}$$

At the melting temperature, $f = \frac{1}{2}$.

$$\frac{1}{T_m} = \frac{R}{\Delta H^0}(n-1)\ln c_t + \frac{\Delta S^0}{\Delta H^0} - \frac{R}{\Delta H^0} \ln \frac{2^{n-1}}{n}. \tag{33a}$$

Of course, there is no dependence on melting temperature for unimolecular reactions ($n = 1$). For these reactions, $T_m = \Delta H^0/\Delta S^0$. For other reactions, the slope of a plot of $1/T_m$ against $\ln c_t$ is proportional to $R/\Delta H^0$. We show eq. (33) as well as eq. (33a) to emphasize that any value of f can be chosen for the characteristic inverse temperature to plot as a function of $\ln c_t$. The slope of the plot is the same.

As an example of a plot of $\ln c_t$ against $1/T_m$, we present fig. 5.4 for the transition of complementary single strands to duplex ($n = 2$). A value of $\Delta H^0 = -20{,}000R$ (J) was used as in fig. 5.3, but ΔS^0 was chosen as $-20{,}000R/353$ (J K^{-1}). This gives a melting temperature of 353 K for standard conditions of 1 M strand concentration, but gives measurable melting temperatures for strand concentrations in the μM to mM range. Note that, for ΔH^0 of about 165 kJ mol^{-1} (40 kcal mol^{-1}), there is roughly a 10 K increase in melting temperature for a factor of ten increase in concentration. Because the value of ΔH^0 depends on the logarithm of the concentration, the absolute magnitude of concentration is irrelevant. Only the relative magnitudes of total strand concentrations affect the value of ΔH^0.

Self-complementary and Non-self-complementary Duplexes

Because so many of the thermodynamic data on nucleic acids have been measured from helix melting, we summarize in table 5.1 the above equations for double-helix formation from self-complementary ($2S \to D$) and non-self-complementary single strands ($S_A + S_B \to D$). The only difference is in how the concentrations of single strands are related to the fraction f of strands in the duplex. This affects the form of the equilibrium constant in terms of f, but it does not affect how f is obtained from the absorbance melting curve, and it does not affect how ΔH^0 depends on the concentration dependence of the melting temperature T_m.

FORMATION OF SECONDARY STRUCTURE IN NUCLEIC ACIDS

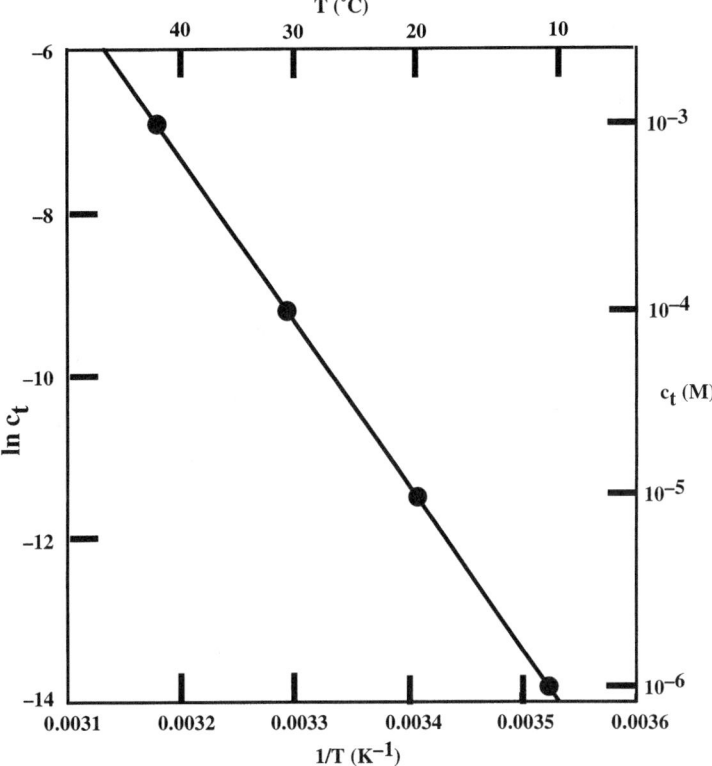

Figure 5.4. A plot of $\ln c_t$ against $1/T_m$ for the reaction $2S \to D$. The slope is $\Delta H^0/R$. The line was calculated for $\Delta H^0 = -20,000R$ (J) and $^0 = -20,000R/353$ (J K^{-1}).

Table 5.1. Thermodynamics of Melting of Double Helicess

Self-complementary $2S \to D$	Non-self-complementary $S_A + S_B \to D$
$[D] = \dfrac{fc_t}{2}$ and $[S] = (1-f)c_t$	$[D] = \dfrac{fc_t}{2}$ and $[S_A] = [S_B] = \dfrac{(1-f)c_t}{2}$
$K = \dfrac{f}{2c_t(1-f)^2}$	$K = \dfrac{2f}{c_t(1-f)^2}$
$\Delta G^0(T_m) = RT_m \ln c_t$	$\Delta G^0(T_m) = RT_m \ln \dfrac{c_t}{4}$
$\Delta H^0(T_m) = 6RT_m^2 \dfrac{\partial f}{\partial T}(T_m)$	$\Delta H^0(T_m) = 6RT_m^2 \dfrac{\partial f}{\partial T}(T_m)$
$\dfrac{1}{T_m} = \dfrac{R}{\Delta H^0} \ln c_t + \dfrac{\Delta S^0}{\Delta H^0}$	$\dfrac{1}{T_m} = \dfrac{R}{\Delta H^0} \ln c_t + \dfrac{\Delta S^0}{\Delta H^0} - \dfrac{R}{\Delta H^0} \ln 4$

Here c_t is the total strand concentration. The non-self-complementary strands are in a 1:1 ratio with each concentration equal to $\tfrac{1}{2} c_t$.

Temperature Dependence of ΔH^0 and ΔS^0

The equations in table 5.1 for $\ln c_t$ against T_m ignore the temperature dependence of ΔH^0 and ΔS^0. For explicit inclusion of the temperature dependence, a ΔC_P term must be added to ΔH^0 and ΔS^0. We can explicitly use temperature-dependent expressions for $\Delta H^0(T)$ and $\Delta S^0(T)$ in eqs. (1) and (2) relating ΔG^0 to $\ln K$.

$$\Delta H^0(T_m) = \Delta H^0(T) + \Delta C_P(T_m - T), \tag{34}$$

$$\Delta S^0(T_m) = \Delta S^0(T) + \Delta C_P \ln \frac{T_m}{T}. \tag{35}$$

Here $\Delta H^0(T)$ and $\Delta S^0(T)$ are the standard enthalpy and entropy at temperature T, and ΔC_P is the difference in heat capacity between products and reactants (assumed independent of T). In principle, $\ln K$ as a function of temperature, or the equivalent $1/T_m$ as a function of $\ln c_t$, could be fitted to ΔH^0, ΔS^0, and ΔC_P at some chosen temperature. However, this is a poor way to obtain ΔC_P and thus the temperature dependence of ΔH^0 and ΔS^0: the limited temperature range of the experiments, plus random errors in the measurement preclude accurate separation of the effects of ΔC_P and ΔH^0 by curve fitting (Chaires, 1997). It is better to use the equations in table 5.1 to obtain ΔH^0 and ΔG^0 in the middle of the temperature range measured. A value of ΔS^0 can then be calculated for this same range. These values should not be relied upon outside the range, unless a ΔC_P value obtained independently is available.

Temperature-jump Measurements

In a temperature-jump experiment, the temperature of the sample is raised a few degrees rapidly, and the absorbance of the sample at the new temperature is measured as a function of time (Eigen and deMayer, 1974). The timescale of measurement is usually in the range of microseconds to milliseconds. The advantage of temperature-jump methods over equilibrium melting curves is the ability to separate changes in absorbance due to minor conformational changes (such as single-strand base stacking) from changes due to the transition of interest (such as hairpin or duplex formation). The timescales for the different processes are very different and are clearly separable in the signal of absorbance against time (see fig. 5.5). For example, with a conventional temperature-jump caused by the discharge of a capacitor across a cell, the unstacking of single strands occurs as fast as the temperature rises (about a microsecond). The magnitude of the time-dependent part of the absorbance signal is a measure of the transition. This can be seen from the temperature derivative of eq. (25):

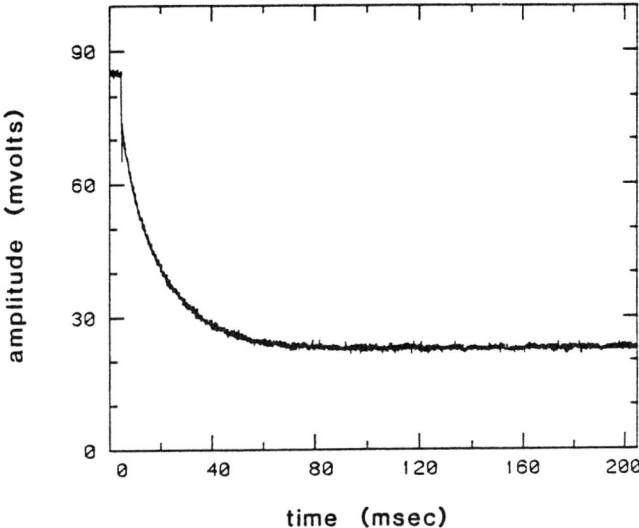

Figure 5.5. An experimental temperature-jump curve for the dissoociation of an RNA oligonucleotide duplex. A 1.8°C temperature jump to a final temperature of 6.7°C was applied to a mixture of rCA$_5$G and rCU$_5$G in solution. The abrupt decrease in signal from 85 to 70 mvolts corresponds to unstacking or other rapid conformational changes. The exponential decay that follows is a measure of the bimolecular reaction. The amplitude of the exponential signal as a function of temperature thus gives a direct measure of df/dT, the derivative of the fraction of duplex, as a function of temperature. No baselines need to be subtracted; the baselines are a measure of the rapid part of the signal.

$$\frac{dA}{dT} = \underbrace{\left[(1-f)\frac{dA_S}{dT} + f\frac{dA_P}{dT}\right]}_{\text{(fast)}} + \underbrace{\left[\frac{df}{dT}(A_P - A_S)\right]}_{\text{(slow)}}. \qquad (36)$$

The first term in brackets represents the change in absorbance of the individual species, due to unstacking or other small conformational changes; it occurs on the nanosecond time scale. The transition of interest is characterized by the second term—the df/dT term. An application to a transition between two RNA secondary structures is given by LeCuyer and Crothers (1994).

Intermediate States

If more than two species and several reactions are involved in the equilibrium, one can write equilibrium constants for each reaction involved. If the concentrations of the different species can be measured, values of ΔG^0, ΔH^0, and ΔS^0 for each reaction can be obtained. For absorbance

146 THERMODYNAMICS IN BIOLOGY

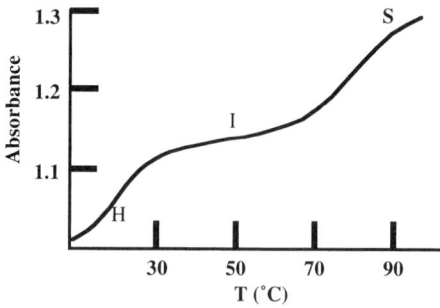

Figure 5.6. An absorbance melting curve for a transition with an intermediate species. The reaction is S → I → H. The single strand to intnermediate equilbrium has $\Delta H^0 = -20,000R$ (J) and $\Delta S^0 = -20,000R/353$ (J) K^{-1}, with $T_m = 80°C$. The intermediate to hairpin equilibrium has $\Delta H^0 = -20,0000R$ (J) and $\Delta S^0 = -20,000R/293$ (J K^{-1}), with $T_m = 20°C$. The temperature dependences of the absorbance of the single strands, the intermediate, and of the hairpins are: $A_S(T) = 1.20 + 0.001(T - 273)$, $A_I(T) = 1.10 + 0.00075(T - 273)$, and $A_H(T) = 1.00 + 0.0005(T - 273)$.

melting curves, the different reactions can only be observed if their T_m values are sufficiently different and the transitions are sharp. Figure 5.6 shows an absorbance melting curve for a reaction with three species: single strands, intermediate, and hairpin.

$$S \to I \to H. \tag{37}$$

The intermediate could be a partially base-paired structure, for example. The two equilibria were chosen with melting temperatures 60 K apart: one at 80°C the other at 20°C. This gives separate transitions with a plateau in between; it allows measurement of the thermodynamic parameters for each equilibrium. Much more often, the transitions will overlap to produce a single broad melting curve, or a melting curve that indicates two transitions, but that cannot be fitted uniquely to two sets of thermodynamic parameters. If a multistate melting curve is fit to a two-state transition, the ΔH^0 value will be smaller than the correct one for the total transition. For consecutive reactions as in eq. (37), the equilibrium constant for the total reaction (S → H) is the product of the two equilibrium constants, and the thermodynamic variables are the sum of the values for the individual reactions. However, the experimental curve will be broader than a two-state curve for the total reaction, and will yield a value of ΔH^0 lower than the sum of the two ΔH^0 values. For reactions that depend on concentration, using a plot of ln c_t against $1/T_m$ can give a better measure of ΔH^0. The melting curves are broad, but their dependence on total concentration is not as directly affected.

If possible, the shape of a melting curve and its concentration dependence should both be used to estimate ΔH^0. Good agreement is an indication of a two-state transition.

For reactions that are independent but that occur in the same temperature range, it may be very difficult to separate and measure their thermodynamic parameters. An example is the noncooperative melting of two domains in a nucleic acid. The domains could have similar transition enthalpies (ΔH_1^0 and ΔH_2^0) and slightly different T_m values. Two-state analysis of the shape of the melting curve could lead to a transition enthalpy less than half the sum of ΔH_1^0 and ΔH_2^0! Less extreme (and less obvious) errors result if the two transitions retain partial cooperative below 1.0. Careful analysis of the total increase of absorbance on melting, of the kinetic properties of the transition, and of the shape of the melting curve would be required to detect this type of non-two-state transition.

Partition Functions

The opposite extreme from a two-state model is a model that includes every species containing one or more base pairs as possible intermediates between single strands and the final structure. A partition function can be constructed to represent the transition, and an attempt can be made to obtain useful thermodynamic parameters. We will describe simple partition functions that apply to the melting of oligonucleotides (Bloomfield et al., 1974; Cantor and Schimmel, 1980). The more general partition functions that are necessary to treat the melting of kilobase pairs of DNA (Poland and Scheraga, 1970) will not be discussed in detail.

As an example, let us consider the transition from a single strand of 14 nucleotides to a hairpin containing five base pairs and a loop of four unpaired nucleotides. The intermediate states can include all species with one to five base pairs. To simplify the model, we will ignore species that have internal loops in the stem, and assume that the stem melts from either end. Only for larger number of base pairs can species with internal loops become significant. We will also ignore base pairs that do not appear in the final structure. Repeating sequences (A_n–U_n) and hundreds of base pairs are needed to lead to significant base pairing in the intermediates not present in the final structure. These assumptions are not of fundamental importance; they can easily be removed. However, it is easy to see that, for helices of less than 10 base pairs, the species omitted will be present in such small concentrations that they are completely negligible.

An equilibrium constant can be written for the formation of each base-paired species from the single strand. The first-base pair forms a loop. There will be a loss of conformational entropy on base-pair formation that depends on the size of the loop. For our model, the loop sizes on forming a base pair can contain 4, 6, 8, 10, or 12 unpaired nucleotides with equilibrium constants $\gamma_4, \gamma_6, \gamma_8, \gamma_{10}$, or γ_{12}. The γ

terms are initiation parameters. Adding a neighboring base pair—a propagation or zippering step—is represented by equilibrium constant s. Therefore all equilibrium constants for the intermediate states have the form $\gamma_n s^j$ ($j = 0, 1, 2, 3, 4$). The final state has equilibrium constant $\gamma_4 s^4$; this is equal to the all-or-none equilibrium constant. The definitions of the initiation and propagation steps are shown in fig. 5.7a: fig. 5.7b shows the possible species present during the transition from single strand to hairpin for the tetraloop with a stem of five base pairs. Each of the γ and s values has a temperature dependence of the form $e^{-\Delta H^0/RT} e^{\Delta S^0/R}$. We expect γ to be less than 1, with a negative ΔS^0 (unfavorable) and a zero or positive ΔH^0 (unfavorable) for initiation. Extra-stable tetraloops (Antao et al., 1991; Antao and Tinoco, 1992) can have a negative ΔH^0, but γ is still less than one. No RNA hairpin loop is known to be stable with only one base pair closing it in aqueous solution (Molinaro and Tinoco, 1995). The value of s, an equilibrium constant for stacking a base pair on one already formed, will depend at least on the nearest neighbor. There are thus ten possible values of s with different values of ΔH^0 and ΔS^0 of propagation corresponding to the ten Watson–Crick nearest neighbors. The partition function of bonded species, q_b, is the sum of all species with at least one base pair. If we assume, for simplicity, equal values of s, then

$$q_b = \sum_{i=0}^{4} \sum_{j=0}^{4-i} \gamma_{2i+4} s^j. \tag{38}$$

The main contributions to the partition function should come from the final state, $\gamma_4 s^4$ (the fully paired tetraloop), plus the two species with one

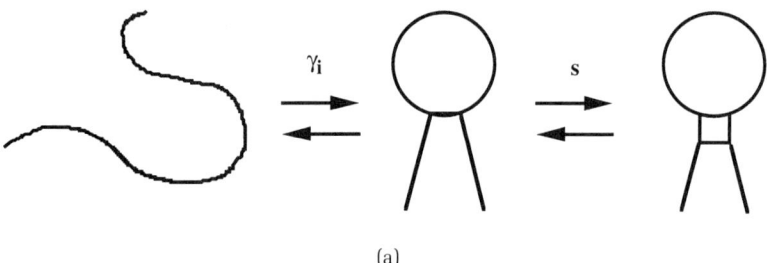

(a)

Figure 5.7. (a) The definition of γ—the equilibrium constant for initiation (forming the first base pair)—and of s: that for propagation (forming the successive base pairs). (b) Schematic of the species involved in forming a hairpin with five base pairs in the stem and a loop of four nucleotides. The equilibrium constant of each species relative to the single strand is shown. For simplicity, all values of s are assumed to be the same.

FORMATION OF SECONDARY STRUCTURE IN NUCLEIC ACIDS 149

(b)

Figure 5.7 (*continued*)

less base pair: $\gamma_4 s^3$ (the tetraloop with the terminal base pair broken) and $\gamma_6 s^3$ (the hexaloop formed by breaking the base pair next to the loop). The fraction f of bonded states is

$$f = \frac{q_b}{1 + q_b}. \tag{39}$$

The absorbance can be written as

$$A = \frac{1}{1+q_b}\left(A_S + \sum_{i=0}^{4}\sum_{j=0}^{4-i} A_{2i+4,j}\gamma_{2i+4}s^j\right), \qquad (40)$$

with A_{nj} the absorbance of each bonded species and A_S the absorbance of the single strand. In general, each value of s will be different; but to extract individual values of ΔH^0, ΔS^0, and $\Delta G^0(T)$ for the loop initiation γ values and the ten possible propagation constants is a daunting task. It requires a combination of many experiments, plus theoretical knowledge of the dependence of initiation constants on loop length. A model of how absorbance depends on the number of base pairs in the different species is also needed.

The partition-function approach to bimolecular reactions gives equations analogous to eqs. (38–40). The initiation equilibrium constant for bringing two strands together is κ. We can expect two different κ values depending on whether initiation occurs at a G–C or at an A–U or A–T base pair. There will be ten possible propagation values, s, corresponding to ten nearest neighbors. We consider short duplexes (less than 20 base pairs) so we can ignore internal loops and strands that pair out of register. That is, we consider only melting from the ends with all the intermediate states on the path to the fully-paired duplex. The partition function for the bonded species of a duplex of N identical base pairs is

$$q_b = \kappa \sum_{j=1}^{N}(N-j+1)s^{j-1}. \qquad (41)$$

There is only one value of κ, and all s values are the same; thus eq. (41) illustrates only the form of the partition function. For a duplex of arbitrary sequence, explicit values for κ and s must be used. Equation (41) then involves a sum of products of the individual values of s (Poland and Scheraga, 1970). The fraction of bonded states for self-complementary strands is obtained from the partition function

$$q_b = \frac{f}{2(1-f)^2 c_t}. \qquad (42)$$

Thus,

$$f = \frac{4q_b c_t + 1 - \sqrt{8q_b c_t + 1}}{4q_b c_t}. \qquad (43)$$

The partition functions provide a formalism for understanding any model of intermediate states that one wants to consider. They can be used to fit melting curves that are clearly not two-state. They can also be used to calculate theoretical melting curves for any assumptions about initiation and propagation parameters. This can show which chain lengths and sequences are inappropriate for two-state models.

FORMATION OF SECONDARY STRUCTURE IN NUCLEIC ACIDS

Individual initiation and propagation parameters can be determined experimentally by studying sequences containing only one or two different nearest-neighbor parameters s.

4. Experimental Methods: Nuclear Magnetic Resonance

Absorbance and circular dichroism measurements monitor concentrations of bonded and unbonded species, but they cannot identify which bases are paired. Nuclear magnetic resonance (NMR) spectra can be assigned to individual nuclei, so in principle each base can be assigned as paired or unpaired.

An NMR spectrum of a mixture of species can be a sum of the spectra of the individual species (slow exchange), or be the average of the species (fast exchange). As the names imply, the difference depends on the rate of exchange between species. For a unimolecular reaction (such as a single strand to hairpin)

$$A \underset{k_{-1}}{\overset{k_1}{\rightleftharpoons}} B,$$

slow exchange occurs when the exchange rate $(k_1 + k_{-1})$ is small compared to the difference in resonant frequencies of the same nucleus in species A and species B. Fast exchange is when $k_1 + k_{-1}$ is large compared to the frequency difference. The different possibilities are summarized in table 5.2.

Table 5.2. NMR Spectrum of a Reacting Mixture of Species: A → B

Slow exchange: $(k_1 + k_{-1})^* < (\nu_A - \nu_B)/10$
Spectrum is sum of spectra of the two species Area of each resonance is proportional to concentration of each species: $\text{area}_A/(\text{area}_A + \text{area}_B) = f_A$; $\text{area}_B/(\text{area}_A + \text{area}_B) = f_B$
Fast exchange: $(k_1 + k_{-1})^* > 10(\nu_A - \nu_B)$
Spectrum is weighted average of spectra of the two species Each resonance frequency is weighted average: $\langle \nu \rangle = f_A \nu_A + f_B \nu_B$
Intermediate exchange: $(k_1 + k_{-1})^* \approx \nu_A - \nu_B$
Spectrum is broadened; each resonance frequency is spread from ν_A to ν_B

ν_A and ν_B are the resonance frequencies of a nucleus in species A and B; f_A and f_B are the fraction of molecules that are present as species A and B.
*For a bimolecular reaction, such as single strands to duplex, 2 A → B, $k_1 + k_{-1}$ is replaced by $k_1[A] + k_{-1}$.

Fast exchange is the most convenient regime to be in, because only one set of resonances is seen—no matter how many species are present. A 1 ppm difference in chemical shifts for a nucleus in different species corresponds to a frequency difference of 500 s^{-1} at 500 MHz, and thus a reaction lifetime of 2 ms. The rates of hairpin formation from a single strand are at least a factor of ten faster; the reaction is in fast exchange. The position of each resonance for a two-state process in fast exchange is

$$\langle v \rangle = f_S v_S + f_{HP} v_{HP}. \tag{44}$$

The observed resonance frequency is $\langle v \rangle$; v_S and v_{HP} are the resonance frequencies of a nucleus in the single strand and in the hairpin; f_S and f_{HP} are the fractions of molecules present as single strand and hairpin. For the range of intermediate states shown in Fig. 7, the average frequency is

$$\langle v \rangle = \frac{1}{1+q_b} \left(v_S + \sum_{i=0}^{4} \sum_{j=0}^{4-i} v_{2i+4,j} \gamma_{2i+4} s^j \right). \tag{45}$$

Here v_S is the resonance frequency of a nucleus in the single strand, and v_{nj} is the resonance frequency of the same nucleus in each of the intermediate species of Fig. 5.7. The partition function of the intermediate states, q_b, is given in eq. (38).

Whether the loop initiation parameters γ_n and the propagation parameters s can be obtained from the curves of $\langle v \rangle$ against temperature will depend on the number of resonances that can be followed through the temperature range. Each nucleotide can be labeled with ^{13}C or ^{15}N isotopes to provide a specific probe for the formation of each base pair.

5. Comparison of Thermodynamics from Equilibrium Constants and from Calorimetry

Values of the enthalpy of a reaction obtained by calorimetry and from the temperature derivative of an equilibrium constant often do not agree. For example, Sturtevant and collaborators (Liu and Sturtevant, 1997) have measured equilibrium constants and heats of reaction by titration of reactants and products in an isothermal calorimeter. The equilibrium constant K and the ΔH value thus come from the same heat effects measured for the same reaction at the same time. The ΔH values from the van't Hoff equation—the temperature dependence of K—and from the heat evolved disagree well beyond any error limits of the measurements. The reasons for this are not known. Sturtevant mentions reactions involving solvent or buffer components that contribute to the heat effects, but that are not explicitly included in the equilibrium reaction (Liu and Sturtevant, 1997). Chaires has analyzed simulated experimental data to illustrate the difficulty in separating the effects of heat capacities and of enthalpies on the temperature dependence of an equilibrium constant (Chaires, 1997). Here we will describe some of the possible contributions to the differences in thermodynamic parameters obtained by different methods.

Thermodynamics is a macroscopic theory that deals with defined components, characterized by chemical potentials and activities relative to chosen reference states. The equations linking the activities to thermodynamic variables of state, such as energy, enthalpy, entropy, free energy, etc. are true essentially by definition. The van't Hoff equation is correct for components and activities. When does it apply to molecules and concentrations? The number of components in a closed system is fixed, but the choice of components is arbitrary. Any linear combination of components is a valid choice. For example, in a system made by dissolving salt-free lyophilized RNA and NaCl in water, there are three components. The properties of the system, such as volume, enthalpy, and entropy depend on the amount of each component, and on the temperature and pressure. We can choose a particular hydrated RNA (e.g. RNA·10H$_2$O) or RNA with bound NaCl (e.g. RNA·10NaCl) as a component (component A) and measure the change of a thermodynamic property of the system when a differential amount, dn_A, of this component is added to the system. The change in volume is related to the partial molal volume $\overline{V}_A$ of component A:

$$dV(\text{system}) = \overline{V}_A \, dn_A. \tag{46}$$

We could as well choose a different number of bound waters or NaCl in defining a component, and measure the partial molal volume of that component. The defined component is not related to the molecular interactions actually occurring in the system. We can define any three components we like, but no more than three, and relate the volume of the system to the sum of the number of moles of the three components:

$$V(\text{system}) = \overline{V}_A n_A + \overline{V}_B n_B + \overline{V}_C n_C, \tag{47}$$

where components A, B, and C are linear combination of RNA, H$_2$O, and NaCl. The values of the partial molal volumes are functions of temperature, pressure, and the concentrations (amounts) of all the components. The same logic and equations apply to partial molal enthalpies, and the enthalpy of the system. We can measure the heat effect at constant pressure, dH, when a differential amount of a component (such as RNA·10 H$_2$O) is added to the system to obtain a value for the partial molal enthalpy $\overline{H}_A$. We could as well choose pure RNA as component A. In either case, we can make measurements in a very dilute aqueous solution of RNA at constant concentration of NaCl to obtain values for $\overline{V}_A$ and $\overline{H}_A$ that depend only on T and P for this essentially constant solvent. For the same conditions, we can obtain $\overline{G}_A$, the partial molal free energy or chemical potential, which depends on the logarithm of concentration c_A of

RNA, and on T, P, and the free energy in the standard state, $\overline{G^0}_A$. The concentration c_A is the concentration of defined component A in the solution; for finite concentrations, it is replaced by the activity a_A.

$$\overline{G}_A = \overline{G^0}_A + RT \ln a_A \approx \overline{G^0}_A + RT \ln c_A. \tag{48}$$

We have chosen the standard state at an RNA concentration of 1 M with the properties of infinitely dilute RNA solution at constant concentration of NaCl. We cannot simply measure a volume change or a heat release in the solution to measure $\overline{G^0}_A$, so the measurement of free energy is more involved. An indirect method is used, such as allowing equilibrium to be established between component A and a component of known standard free energy. At equilibrium at constant T and P, the free energies of the two components are equal, so the difference in standard free energies is obtained from the concentrations (activities) at equilibrium. A pH meter is a common example of this method of measuring activities. The standard free-energy change for a reaction obtained from the ratio of equilibrium concentrations in a solution dilute in reactants and products is another application of the method. The standard free-energy change obviously depends on the definition of the standard state. For each different NaCl concentration, as well as each different T and P, a different standard free energy will be obtained. Of course, the component must be specified for which the free energy is being measured.

How do we interpret changes in the thermodynamic properties of component A (RNA or RNA–ligand) as the temperature is varied at constant P? The thermodynamics is simple: the change in enthalpy is the heat capacity; the change in free energy is a measure of the entropy and the enthalpy:

$$\left[\frac{\partial \overline{H}}{\partial T}\right]_P = \overline{C}_P, \quad \left[\frac{\partial \overline{G}}{\partial T}\right]_P = -\overline{S}, \quad \left[\frac{\partial (\overline{G}/T)}{\partial (1/T)}\right]_P = \overline{H}. \tag{49}$$

However, if we think that the RNA component is undergoing a change in conformation—a hairpin to a single strand—as the temperature is raised over a narrow range, we can attempt to separate 'excess' heat capacities and enthalpies from the background temperature-dependent values if no conformational change occurred. That is, we try to separate the effects of temperature on the hairpin and the single strand from the effects of the transformation form one to the other. We have left the realm of pure thermodynamics and made molecular interpretations. We assume base lines in scanning calorimetry, and assume concentrations of species based on optical properties, for example, in equilibrium experiments. It is not too surprising if the results do not agree.

To remain closer to thermodynamics, we can add a fourth component (a complementary RNA) to the system, and measure changes in the thermodynamic properties of the system as the amounts of components are varied. This is what is done in a titration experiment in an isothermal calorimeter to obtain a ΔH. To obtain a standard free energy for the reaction, ΔG^0, we need to know the activities—assumed equal to concentrations in very dilute solution—of the two reacting components at equilibrium. The van't Hoff ΔH^0 is obtained from the temperature dependence of ΔG^0 from eq. (49). This is similar to measuring the melting temperature T_m of a duplex as a function of the concentration of the strands. The thermodynamic equations state that ΔH^0 obtained by the two methods are equal. What contributes to their experimental differences?

When two strands react, ions and solvent molecules will be released or bound. A reaction can be written explicitly as

$$S_A + S_B \rightarrow D + \Delta n_{Na} Na^+ + \Delta n_w H_2O + \Delta n_H H^+ + \cdots,$$

where the values of Δn_{Na}, Δn_w, and Δn_H can be positive or negative and need not be integers. The changes in bound ligands can contribute to the heat effects, of course. From the heat evolved as component A and B are mixed in an isothermal calorimeter, ΔH is obtained for whatever reaction occurs. Titration of reactants can assure that the reaction is complete. The heat can be measured in dilute enough solution, at constant NaCl concentration, pH, and so on, so that a ΔH^0 is obtained for the defined standard state. It is important to emphasize that the enthalpy change measured is for our *defined* components reacting under the standard state conditions. We do not have to know about the ligands bound or released. We can also measure the concentration of each component in the solution from absorbance measurements on the individual strands in the chosen solvent. We could not do this with the reaction of hairpin to single strand. By mixing the components in dilute solution in the standard state solvent, and measuring absorbance (or NMR), we can measure the equilibrium concentrations of components A and B, and thus D. From this equilibrium constant, we obtain a standard free energy ΔG^0 for the reaction with whatever number of ions and solvent are released or bound. If we wanted to determine ΔH^0 and ΔG^0 for the unliganded strands, we would need to measure in separate experiments their binding of Na^+ ions, H^+ ions, and so on. A good example of the experimental data needed for this type of analysis is given for the reaction of two ATP molecules to form ADP and AMP (Goldberg and Tewari, 1991; Tewari and Goldberg, 1991).

We have shown how standard thermodynamic variables can be obtained for a reaction between two components that can be added to a constant solvent at constant T and P. We measure the heat effects to obtain ΔH^0, and we measure any property (absorbance, NMR, fluorescence, etc.) directly related to amount of component to obtain an equilibrium constant and ΔG^0. It is important that the components be dilute and that the salt concentration, pH, and so on of the solvent remain constant independent of the reaction taking place. We thus obtain ΔG^0, ΔH^0, and (from eq. (1) ΔS^0 at temperature T (and 1 atm pressure).

Obtaining ΔH^0 amd ΔS^0 from the temperature dependence of ΔG^0 (see eq. 49) is more complicated. If both ΔH^0 and ΔS^0 are independent of temperature, the simple van't Hoff equation should be a good approximation. However, we expect both ΔH^0 and ΔS^0 to change with temperature, because of intrinsic changes in A and B and in the number of ligands released. This means that ΔC_P is not zero, and must be included in the temperature dependence of ΔG^0. Equations (34, 35) give the temperature dependence of ΔH^0 and ΔS^0 when ΔC_P is independent of T. Since separating ΔH^0 and ΔC_P from temperature dependence of $\ln K$ or $\Delta G^0/T$ is very difficult (Chaires, 1997), this may help explain the differences in ΔH^0 values from calorimetry and van't Hoff.

It is important to notice that obtaining thermodynamic parameters for the reaction between two strands is not changed if one of the strands is in equilibrium with a hairpin conformation. As long as the absorbance has been calibrated for this component—the equilibrium mixture of single strand and hairpin—an equilibrium constant and ΔG^0 is obtained for the reaction between the two strands. The temperature dependence of $\ln K$ will have an extra large ΔC_P, and this ΔC_P will depend on temperature, but otherwise there is no change in the analysis of the data. That is, a two-state analysis can be correctly applied to a multi-state reaction. As long as there is a reaction of two components with known stoichiometry, and the equilibrium concentrations are measured, a two-state analysis is valid.

Clearly, more experiments are needed to establish when a set of measured data can be given a rigorous thermodynamic analysis.

6. The Additivity of Thermodynamic Data

Thermodynamic data have been measured for many different transitions in nucleic acids. It is important to be able to systematize these data so that the results can be extrapolated, and can be used to estimate thermodynamics of other transitions. If we assume that the folded conformation of a nucleic acid is one that minimizes the Gibbs free energy of the system, then the thermodynamic data can be used to predict the conformation of a folded state.

The basic assumption in understanding the thermodynamics of nucleic acids is that ΔG^0 values—and ΔH^0 and ΔS^0—are additive for the structural elements in the folded state. The thermodynamic parameters are measured relative to the single-strand state. The largest amount of data is, of course, for double helices of Watson–Crick base pairs. A single strand folds into a collection of Watson–Crick double helices, hairpin loops, internal loops, bulges, junctions, and tertiary interactions. There is no agreement in the literature on the distinction between secondary structures and tertiary structures. The following definition is most useful for prediction of secondary structure from thermodynamic data. The sequence of the nucleic acid is drawn in a plane with the backbone forming a continuous closed boundary when the 5' and 3' ends are joined. Watson–Crick pairing between the bases in the sequence are depicted by lines joining the bases; the lines must remain within the closed boundary. A secondary structure has no lines crossing, as shown in fig. 5.8, which depicts the different types of secondary structural elements. Pairing bases within the loops, bulge or junction of fig. 5.8 is allowed as a part of a secondary structure, but pairing a base betwen any two secondary structural elements would require crossing of lines, and would be classified as a tertiary structure. Base triplets—hydrogen bonding between a helix and a loop, bulge, or junction nucleotide—are tertiary structure. The reason that prohibition of so-called chord crossing is a useful criterion for defining secondary structure is that the method of dynamic programming (Bellman, 1958; Nussinov et al., 1978) can be used to find a global optimum for structures that do not have chord crossing. A chord representing a Watson–Crick base pair between any two bases separates the secondary structure into two parts whose energies can be calculated independently. This means that all possible secondary-structure base pairs can be considered, and the free energies of the structures calculated. The global minimum can be found. Similarly, terms in the partition function restricted to the corresponding parts of the molecule can be calculated, and the total partition function can be determined (McCaskill, 1990). The dynamic-programming methods need consider at most terms of the order of N^3, where N is the number of nucleotides. If interaction were allowed between the bulge and the internal loop of fig. 5.8, for example, then these interactions would no longer allow the separation of the structure into independent units. The number of possibilities to consider would increase exponentially with the number of nucleotides. Dynamic programming could not be used, and an exhaustive search for a global minimum would require exponentially increasing time.

The free energy of a secondary structure is calculated as the sum of free energies of its constituent elements. Double-strand helices lower the free energy; loops, bulges, and junctions raise the free energy. Clearly, the more data available for each type of element, the more accurate the calculated free energy. The strategy is to measure thermodynamic parameters for nucleic acids that contain the different elements, then to divide the measured parameters into contributions from the elements. Double helices have been studied the most, because they are the only secondary structural elements that can occur alone. Their thermodynamics can be represented as a sum of nearest-neighbor base pairs. This will be discussed in the next section.

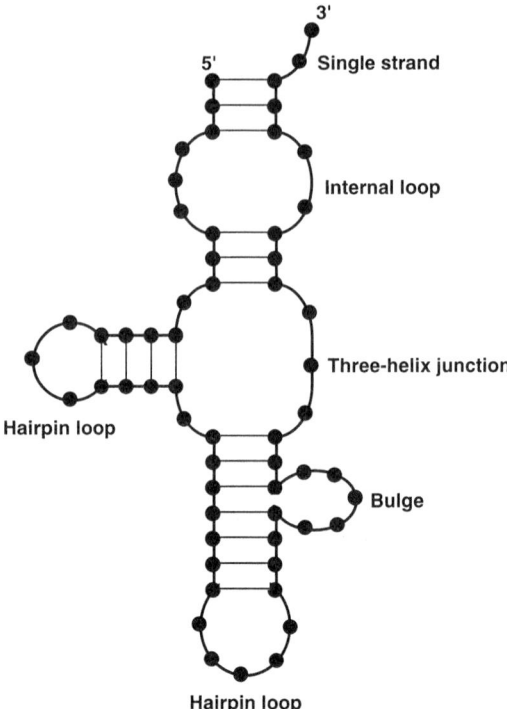

Figure 5.8. A diagram of a secondary structure illustrating the different structural motifs. Each chord represents a base pair. A tertiary-structure motif, such as a base pair between two loops, would require the crossing of two chords.

Thermodynamic parameters of hairpin loops depend on their lengths, their sequences, and their closing base pairs. Some tetraloop sequences are exceptionally stable (Antao et al., 1991). The larger the loop, the more destabilizing it is, because of the loss of entropy relative to the single strand. However, hairpin loops of more than about ten nucleotides are rare in nature, because base pairing within the loop can occur to form an internal loop and a smaller hairpin. At higher temperatures, large hairpin loops can be observed; loop sizes of up to about 80 nucleotides have been observed by temperature-jump experiments on the denaturation process of viroid RNAs (Henco et al., 1979).

Internal loops are defined as interruptions of helices by non-Watson–Crick base appositions. A single base–base mismatch is considered an internal loop of two—even though hydrogen bonding occurs, as in G–U and G–A mismatches. The free energy of the molecule can even be lowered by a two-base internal loop, as found in G–U mismatches (He et al., 1991). Isolated G–U and G–T mismatches are consistent with the nearest-neighbor model and might well be considered base pairs. The thermodynamic parameters of internal loops depend on the number and sequence of the nucleotides, whether the loop is symmetric or asymmetric, and on the closing base pairs.

Bulges interrupt helices only on one of the two strands. The other strand has continuous Watson–Crick base pairing. A one-base bulge can be either intercalated or not, but the larger bulges will necessarily be extrahelical. Thermodynamic parameters for bulges can be obtained by assuming that base-pair stacking is continuous for a one-nucleotide bulge, but that stacking of the adjacent base pairs is disrupted by two or more nucleotides in the bulge (Jaeger et al., 1989). However, general consensus has not been reached on helix stacking on either side of a bulge.

Junctions have the most variables affecting their thermodynamics. There are junctions of three, four, or more helices; a two-helix junction is an internal loop. There can be zero, one, two, or more nucleotides between each pair of helices. Finally, the closing pairs on all the helices in a junction are important.

Double-strand Nearest-neighbor Parameters

Data on thermodynamics of double-strand formation by two single strands have been analyzed in terms of nearest-neighbor interactions (Borer et al., 1974). The measured thermodynamic parameters are written as a sum of an initiation contribution (I) plus ten Watson–Crick nearest-neighbor contributions:

5'AA3' 5'AC3' 5'AG3' 5'AT3' 5'CA3'   5'CC3' 5'CG3' 5'GA3' 5'GC3' 5'TA3'
3'TT5' 3'TG5' 3'TC5' 3'TA5' 3'GT5'   3'GG5' 3'GC5' 3'CT5' 3'CG5' 3'AT5'.

For example, a measured thermodynamic parameter for a forming a 5-base-pair helix from its single strands

$$5'ACTGG3'$$
$$3'TGACC5'$$

is

$$I+5'AC + CT + TG + GG3'$$
$$3'TG + GA + AC + CC5'.$$

The initiation parameter takes into account the entropy loss on bringing the two strands together to form the duplex, but it may also contain an enthalpy contribution. An overdetermined set of oligonucleotides is studied to find the best least-squares values of the parameters. This model clearly ignores end effects. The contribution of a particular nearest neighbor is assumed to be the same, whether or not it is on an end. The model has been extended by considering more than nearest neighbors (Gray and Tinoco, 1970) and by adding end effects (Goldstein and Benight, 1992;

Gray, 1997 a,b). However, the prediction of secondary structure so far has used the simple nearest-neighbor-only model.

7. Experimental Results

Double-helix Formation

Thermodynamic data for double-helix formation of Watson–Crick base pairs in RNA (Borer et al., 1974; Pörschke et al., 1973; Turner et al., 1988) and DNA (Gotoh, 1983; Breslauer et al., 1986; Klump, 1990; SantaLucia et al., 1996) have been available for many years. More recently, data for DNA–RNA hybrids were published (Sugimoto et al., 1995). The nearest-neighbor helix propagation parameters and initiation parameters for double-helix formation of Watson–Crick base pairs are given in tables 5.3–5.5. The accuracy of stacking stability parameters is suitable for prediction of oligonucleotide duplex stability as well as for longer duplexes.

The process of oligonucleotide duplex formation depends on concentration. This is taken into account by an equilibrium constant β for initiation of double-strand formation, and by the corresponding initiation enthalpy, entropy, and free energy:

$$\beta = e^{-\Delta H_i^0/RT} e^{\Delta S_i^0/R} = e^{-\Delta G_i^0/RT}. \tag{50}$$

The main contribution to β comes from the loss of entropy on bringing the two strands together, so the standard initiation enthalpy ΔH_{init}^0 is often set to zero. The standard initiation parameters refers to 1 M strand concentration. To calculate thermodynamic properties, such as melting temperatures, at arbitrary strand concentrations, we use

$$S_{\text{init}} = S_{\text{init}}^0 + R \ln c, \qquad H_{\text{init}} = H_{\text{init}}^0. \tag{51}$$

The standard initiation parameters for double-strand formation are given at the bottom of tables 5.3–5.5 for Watson–Crick nearest-neighbor parameters.

Self-complementary strands carry an additional symmetry-correction entropy ΔS_{sym}^0, and corresponding $\Delta G_{37\,\text{sym}}^0$, that must be added to account for the fact that two strands of the same species associate to form a duplex. The number of possible sites for initiation are half the number for a non-self-complementary pair of strands:

$$\begin{aligned}\Delta S_{\text{sym}}^0 &= -R \ln 2 = -1.38\,\text{cal mol}^{-1}\,\text{K}^{-1} = -5.76\,\text{J mol}^{-1}\,\text{K}^{-1},\\ \Delta G_{37\,\text{sym}}^0 &= 0.43\,\text{cal mol}^{-1} = 1.78\,\text{J mol}^{-1}.\end{aligned} \tag{52}$$

Table 5.3. Nearest-neighbor Parameters for Watson–Crick Base Pairs in RNA in 1 M NaCl at pH 7. Data are from Turner et al. (1988). The precision of the values are in the range of 5%; extra significant figures are provided for optimum calculations of T_m and for calculations of ΔG^0 at other temperatures.

Stack type	ΔH^0		ΔS^0		ΔG^0_{37}	
	kJ mol^{-1}	kcal mol^{-1}	J mol^{-1} K^{-1}	cal mol^{-1} K^{-1}	kJ mol^{-1}	kcal mol^{-1}
A–U/A–U						
5'AA3' 3'UU5'	−27.54	−6.58	−76.84	−18.35	−3.71	−0.89
5'AU3' 3'UA5'	−23.97	−5.72	−64.99	−15.52	−3.81	−0.91
3'UA5' 5'AU3'	−34.01	−8.12	−94.45	−22.56	−4.71	−1.13
A–U/G–C						
5'CA3' 3'GU5'	−43.80	−10.46	−116.55	−27.84	−7.65	−1.83
5'CU3' 3'GA5'	−31.96	−7.63	−80.51	−19.23	−6.99	−1.67
5'GA3' 3'CU5'	−55.61	−13.28	−148.51	−35.47	−9.55	−2.28
5'GU3' 3'CA5'	−42.90	−10.25	−109.76	−26.22	−8.86	−2.12
G–C/G–C						
5'CG3' 3'GC5'	−33.58	−8.02	−81.08	−19.37	−8.44	−2.02
5'GC3' 3'CG5'	−59.52	−14.22	−146.17	−34.91	−14.18	−3.39
5'GG3' 3'CC5'	−50.92	−12.16	−124.55	−29.75	−12.28	−2.93

ΔH^0(initiation) = 0; ΔS^0(initiation) = −45.33 J mol^{-1} K^{-1} = −10.83 cal mol^{-1} K^{-1}; ΔG^0_{37}(initiation) = 14.06 kJ mol^{-1} = 3.36 kcal mol^{-1}.

Thermodynamic data for stacking of bases on the ends of RNA helices—the effects of so-called dangling ends (Turner et al., 1988)—and for coaxial stacking of helices (Walter et al., 1994; Walter and Turner, 1994) are also available. The sequence dependence of these stacking interactions is consistent with the nearest-neighbor stacking interactions given in table 5.3.

Calculation of the full partition function with a recursive algorithm (Poland, 1974) is required to describe thermal stability of long (hundreds of base pairs) double strands. This is necessary, because intermediate states involving internal loop formation, partial melting from the ends, and so on become significant for the long double strands. A partition-function analysis may also be required for shorter sequences, if they do not melt in a two-state process.

Table 5.4. Nearest-neighbor Parameters for Watson–Crick Base Pairs in DNA in 1 M NaCl at pH 7. Data are from Allawi and SantaLucia (1997). The precision of the values is in the range of 5–10%; extra significant figures are provided for optimum calculations of T_m and for calculations of ΔG^0 at other temperatures.

Stack type	ΔH^0		ΔS^0		ΔG^0_{37}	
	kJ mol^{-1}	kcal mol^{-1}	J mol^{-1} K^{-1}	cal mol^{-1} K^{-1}	kJ mol^{-1}	kcal mol^{-1}
A–T/A–T						
5'AA3' 3'TT5'	−33.1	−7.9	−93.0	−22.2	−4.19	−1.00
5'AT3' 3'TA5'	−30.2	−7.2	−85.4	−20.4	−3.68	−0.88
5'TA3' 5'AT5'	−30.2	−7.2	−89.2	−21.3	−2.43	−0.58
A–T/G–C						
5'CA3' 3'GT5'	−35.6	−8.5	−95.0	−22.7	−6.07	−1.45
5'CT3' 3'GA5'	−32.7	−7.8	−87.9	−21.0	−5.36	−1.28
5'GA3' 3'CT5'	−34.3	−8.2	−93.0	−22.2	−5.44	−1.30
5'GT3' 3'CA5'	−35.2	−8.4	−93.8	−22.4	−6.03	−1.44
G–C/G–C						
5'CG3' 3'GC5'	−44.4	−10.6	−113.9	−27.2	−9.09	−2.17
5'GC3' 3'CG5'	−41.0	−9.8	−102.2	−24.4	−9.34	−2.24
5'GG3' 3'CC5'	−33.5	−8.0	−83.3	−19.9	−7.70	−1.84

ΔH^0(G–C initiation) = 0.4 J mol^{-1} = 0.1 cal mol^{-1}; ΔH^0(A–T initiation) = 0.6 J mol^{-1} = 2.3 cal mol^{-1}; ΔS^0 (G–C initiation) = −11.7 J mol^{-1} K^{-1} = −2.8 cal mol^{-1} k^{-1}; ΔS^0(A–T initiation) = 17.2 J mol^{-1} K^{-1} = 4.1 cal mol^{-1} k^{-1}; ΔG^0_{37}(G–C initiation) = 4.10 kJ mol^{-1} = 0.98 kcal mol^{-1}; ΔG^0_{37}(A–T initiation) = 4.31 kJ mol^{-1} = 1.03 kcal mol^{-1}.

Dependence on Ionic Strength

The thermodynamics of double-strand formation depends on the chemical nature and the concentration of the salts in the solution. The most extensive measurements (tables 5.3–5.5) have been made in 1 M NaCl, but data for DNA duplex formation have also been measured at 100 mM NaCl (Klump, 1990) and 19 mM NaCl (Gotoh, 1983).

For a reaction involving ions, the concentrations of the ions will affect the equilibrium constant, and the melting temperature. For a reaction in which metal ions M are bound (or released), the equation can be written as

Table 5.5. Nearest-neighbor Parameters for Watson–Crick Base Pairs in DNA–RNA hybrids in 1 M NaCl at pH 7. Data are from Sugimoto et al. (1995). The precision of the values is in the range of 5–10%; extra significant figures are provided for optimum calculations of T_m and for calculations of ΔG^0 at other temperatures.

Stack type	ΔH^0		ΔS^0		ΔG^0_{37}	
	kJ mol^{-1}	kcal mol^{-1}	J mol^{-1} K^{-1}	cal mol^{-1} K^{-1}	kJ mol^{-1}	kcal mol^{-1}
A–T/A–U						
5'rAA3' 3'dTT5'	−32.7	−7.8	−91.7	−21.9	−4.2	−1.0
5'rAU3' 3'dTA5'	−34.8	−8.3	−100.1	−23.9	−3.8	−0.9
3'rUA5' 5'dAT3'	−32.7	−7.8	−97.1	−23.2	−2.5	−0.6
3'rUU5' 5'dAA3'	−48.1	−11.5	−152.4	−36.4	−0.8	−0.2
A–T/G–C						
5r'CA3' 3'dGT5'	−37.7	−9.0	−109.3	−26.1	−3.8	−0.9
5'rCU3' 3'dGA5	−29.3	−7.0	−82.5	−19.7	−3.8	−0.9
5'rGA3' 3'dCT5'	−23.0	−5.5	−56.5	−13.5	−5.4	−1.3
5'rGU3' 3'dCA5'	−32.7	−7.8	−90.4	−21.6	−4.6	−1.1
5'rUG3' 3'dAC5'	−43.5	−10.4	−118.9	−28.4	−6.7	−1.6
5'rUC3' 3'dAG5'	−36.0	−8.6	−95.9	−22.9	−6.3	−1.5
5'rAG3' 3'dTC5'	−38.1	−9.1	−98.4	−23.5	−7.5	−1.8
5'rAC3' 3'dTG5'	−24.7	−5.9	−51.5	−12.3	−8.8	−2.1
G–C/G–C						
5'rCG3' 3'dGC5'	−68.2	−16.3	−197.2	−47.1	−7.1	−1.7
5'rGC3' 3'dCG5'	−33.5	−8.0	−71.59	−17.1	−11.3	−2.7
5'rGG3' 3'dCC5'	−53.6	−12.8	−133.6	−31.9	−12.1	−2.9
5'rCC3' 3'dGG5'	−38.9	−9.3	−97.1	−23.2	−8.8	−2.1

ΔH^0(initiation) = −7.95 kJ mol^{-1} = −1.9 kcal mol^{-1}; ΔS^0(initiation) = −16.2 J mol^{-1} K^{-1} = −3.9 cal mol^{-1} K^{-1}; ΔG^0_{37}(initiation) = 12.98 kJ mol^{-1} = 3.1 kcal mol^{-1}

164 THERMODYNAMICS IN BIOLOGY

$$D \to S_A + S_B + \Delta n M \tag{53}$$

with Δn a positive number (when ions are preferentially bound by the duplex, and are released on melting the duplex) or negative number (when ions are preferentially bound to the single strands). The equilibrium constant can be written explicitly including the metal ion, so that K depends only on T and P. Note that the equation is written as a dissociation of the duplex, instead of its formation as we have been previously writing it:

$$K(T, P) = \frac{[S_A][S_B][M]^{\Delta n}}{[D]}. \tag{54}$$

The equilibrium constant can be written in terms of a K_{obs} that implicitly includes the effect of the metal ion:

$$K(T, P) = K_{obs}[M]^{\Delta n}, \qquad K_{obs} = (T, P, [M]) = \frac{[S_A][S_B]}{[D]}. \tag{55}$$

Since $K(T, P)$ is independent of Δn ($\partial \ln K / \partial \ln [M] = 0$), it follows that

$$\frac{\partial \ln K_{obs}}{\partial \ln [M]} = -\Delta n. \tag{56}$$

This equation describes the fact that, if metal ions are released on melting the duplex (Δn positive), then increasing concentration of M will increase the concentration of duplex—the K_{obs} will decrease. Similarly, if metal ions are preferentially bound by the single strands (Δn negative), increasing [M] will increase K_{obs} and the concentration of single strands.

The K values have the usual temperature dependence:

$$\frac{\partial \ln K}{\partial T} = \frac{\Delta H^0}{RT^2} \quad \text{and} \quad \frac{\partial \ln K}{\partial (1/T)} = -\frac{\Delta H^0}{R}. \tag{57}$$

Combining eqs. (56) and (57), we get for the dependence of melting temperature T_m on metal-ion concentration:

$$\frac{\partial T_m}{\partial \ln[M]} = \frac{\Delta n R T_m^2}{\Delta H^0} \quad \text{and} \quad \frac{\partial (1/T_m)}{\partial \ln[M]} = \frac{-\Delta n R}{\Delta H^0}. \tag{58}$$

Equation (58) is a general equation for the effect of a ligand on a melting temperature. The ligand can raise or lower the melting temperature, depending on the sign of Δn and the sign of ΔH^0. If ligand is preferentially bound by the double helix (Δn is positive) and heat is absorbed on melting (ΔH^0 is positive), increasing concentration of ligand favors the double helix and raises its melting temperature.

Melting temperatures for DNA and RNA in 1 M NaCl can be calculated from the strand concentration, the ΔH^0 and ΔS^0 values in Tables 5.3–5.5, and the equations given in Table 5.1. At other ionic strengths, the T_m values can be estimated by empirical equations that relate the melting temperature to the logarithm of the concentration of Na^+.

The following equation can be used to extrapolate melting temperatures to a Na$^+$ concentration different from 1 M:

$$\Delta T_m = (f_{GC} I_{GC} + f_{AT} I_{AT}) \log M, \qquad (59)$$

where ΔT_m is the difference in melting temperatures, M is the concentration of Na$^+$, f_{GC} and f_{AT} are the fraction of GC and AT (AU) base pairs, respectively, and I_{GC} and I_{AT} are respectively the ionic-strength dependence of GC and AT (AU) base pairs. For DNA, the ionic-strength dependence of base pairs is (Frank-Kamenetskii, 1971; Owen et al., 1969) $I_{AT} = 18.3°C$ and $I_{GC} = 11.3°C$. For RNA, the ionic-strength dependence of base pairs is (Steger et al., 1980) $I_{AU} = 20.0°C$ and $I_{GC} = 8.4°C$.

Mismatches

There are 48 possible nearest-neighbor sequences of isolated base–base mismatches surrounded by Watson–Crick base pairs. Only a small fraction of the possibilities has been studied. In DNA, the series dCA$_3$XA$_3$G + dCT$_3$YT$_3$G ($X, Y = A, C, G, T$) was investigated (Aboul-ela et al., 1985). The G–T, G–G, and G–A mismatches were most stable; the C–A and C–C mismatches were least stable. A similar order of stabilities was found for the DNA duplex

dCGTCGTTTX + dCGACGTTGTYAAACGACG

(Werntges et al., 1986). Nearest-neighbor thermodynamic data for single G–U (He et al., 1991) and G–T base pairs (Allawi and Santalucia, 1997) in Watson–Crick duplexes are given in tables 5.6–5.7. However, significant non-nearest-neighbor effects have been reported for tandem G–U pairs (He et al., 1991). In RNA, symmetric tandem mismatches of the type

$$5'XY3'$$
$$3'YX5'$$

surrounded by both A–U and G–C base pairs were studied (He et al., 1991; Wu et al., 1995). The order of decreasing stability is

5'UG3' 5'GU3' 5'GA3' 5'AG3' 5'UU3' 5'CA3'   5'CU3' 5'UC3' 5'CC3' 5'AC3' 5'AA3'
3'GU5' 3'UG5' 3'AG5' 3'GA5' 3'UU5' 3'AC5'   3'UC5' 3'CU5' 3'CC5' 3'CA5' 3'AA5.

The mismatchese destabilize the helix less when surrounded by G–C base pairs than by A–U base pairs. Significant non-nearest-neighbor effects occur.

Loops and Bulges

Isolated base–base mismatches are symmetric internal loops of two, and tandem mismatches are symmetric internal loops of four. In general, internal loops can have many nucleotides (two or more) and be asymmetric as well as symmetric. Other loop types occurring in RNA and DNA second-

Table 5.6. Nearest-neighbor Parameters for G–U Base Pairs in RNA in 1 M NaCl at pH 7. Data are from He et al. (1991).

Stack type	ΔH^0		ΔS^0		ΔG^0_{37}	
	kJ mol^{-1}	kcal mol^{-1}	J mol^{-1} K^{-1}	cal mol^{-1} K^{-1}	kJ mol^{-1}	kcal mol^{-1}
A–U/G–U						
5'AG3' 3'UU5'	−11.7	−2.8	−30.1	−7.2	−2.1	−0.5
5'AU3' 3'UG5'	−16.8	−4.0	−40.6	−9.7	−4.2	−1.0
5'GA3' 3'UU5'	−56.9	−13.6	−168.3	−40.2	−4.6	−1.1
5'UG3' 3'AU5'	−35.6	−8.5	−104.2	−24.9	−3.4	−0.8
G–C/G–U						
5'CG3' 3'GU5'	−13.0	−3.1	−26.0	−6.2	−5.0	−1.2
5'CU3' 3'GG5'	−46.9	−11.2	−126.0	−30.1	−8.0	−1.9
5'GC3' 3'UG5'	−39.8	−9.5	−100.1	−23.9	−8.8	−2.1
5'GG3 3'CU5'	−26.4	−6.3	−66.2	−15.8	−5.9	−1.4

ary structure are bulge loops (or bulges), hairpin loops, and multibranch loops (or junctions). Experimental data characterizing loop stability are not extensive, and simple models based on the assumption that the destabilizing effect of a loop is purely entropic have been found to be oversimplified. There is need for much more experimental data.

Hairpin loops are among the best studied, with fewer experimental data available for internal and bulge loops. Based on comparative sequence (phylogenetic) analysis of ribosomal RNA, a class of hairpin loops with four unpaired nucleotides was found to be overrepresented in rRNA structures; subsequent experiments revealed both exceptional stability and unusual loop conformations for several of these tetraloops (Varani, 1995). Similar sequence analysis led to investigations on small symmetric internal loops, again supporting a correlation between stability and frequency of occurrence in ribosomal RNA. A periodic table of energies for these small loops has been proposed, based on experimental data for palindromic sequences and extrapolation to the general case (Wu et al., 1995).

Early measurements on a small set of loops established the principal rules for size dependence of the loop entropy, as reviewed by Turner et al. (1988). Further studies have recently been done on hairpin loops of three,

Table 5.7. Nearest-neighbor Parameters for G–T Base Pairs in DNA in 1 M NaCl at pH 7. Data are from Allawi and SantaLucia (1997).

Stack type	ΔH^0		ΔS^0		ΔG^0_{37}	
	kJ mol^{-1}	kcal mol^{-1}	J mol^{-1} K^{-1}	cal mol^{-1} K^{-1}	kJ mol^{-1}	kcal mol^{-1}
A–T/G–T						
5'AG3' 3'TT5'	1.26	0.3	−5.86	−1.4	3.01	0.72
5'AT3' 3'TG5'	−12.56	−3.0	−41.87	−10.0	0.38	0.09
5'GA3' 3'TT5'	−4.61	−1.1	−18.84	−4.5	1.30	0.31
5'TG3' 3'AT5'	−0.42	−0.1	−6.70	−1.6	1.67	0.40
G–C/G–T						
5'CG3' 3'GT5'	−17.17	−4.1	−49.40	−11.8	−1.84	−0.44
5'CT3' 3'GG5'	−15.49	−3.7	−46.05	−11.0	−1.26	−0.30
5'GC3' 3'TG5'	−20.93	−5.0	−59.45	−14.2	−2.47	−0.59
5'GG3 3'CT5'	10.05	2.4	32.24	7.7	0.08	0.02

four, and six nucleotides (Serra et al., 1993, 1994, 1997), and on internal loops of 3 and 4 nucleotides (Schroeder et al., 1996), to establish the sequence dependence of loop thermodynamics. Results from these studies suggest that the main contribution to sequence-specific effects can be accounted for by means of "terminal mismatches". The bases immediately adjacent to the closing base pairs are assumed to stack on top of the closing base pairs, thus causing both enthalpic and entropic contributions to the loop energy. This model had been proposed earlier (Jaeger et al., 1989) based on data for dangling-end stacking and terminal mismatches derived from oligonucleotide duplex stabilities (Turner et al., 1988). Asymmetric internal loops are assumed to be less stable than symmetric ones. Jaeger et al. (1989) reduced the asymmetry penalties introduced by Papanicoplaou et al. (1984) by a factor of 2 to improve structure prediction. However, more recent studies (Peritz et al., 1991) have found these penalties to be still too large in general—and, of course, sequence effects are very important.

For bulge loops, there can be a stabilizing contribution from stacking between the base pairs closing the loop (Longfellow et al., 1990). However, there is no general agreement on whether to consider all bulge loops as stacked across the loop regardless of size, or to restrict this additional contribution to small bulge loops. Jaeger et al. (1989) assume that the closing base pairs are stacked only for one-base bulges.

Table 8.5 summarizes internal, bulge, and hairpin loop entropies for loops of one to ten nucleotides (Jaeger et al., 1989). Entropies for larger loops are extrapolated according to a theoretical model based on end-to-end distances of Gaussian chains (Jacobson and Stockmayer, 1950):

$$\Delta S^0(n) = \Delta S^0(n_{\exp}) - 1.75R \ln(n/n_{\exp}). \tag{60}$$

Here $\Delta S^0(n)$ is the entropy change for forming a loop of n nucleotides, and $\Delta S^0(n_{\exp})$ is the experimentally measured entropy change for forming a loop of $n_{\exp}$ nucleotides. Another extrapolation based on a similar model (Inners and Felsenfeld, 1970) uses

$$\Delta S^0(n) = R \ln[0.000957(n+1)^{-1.5} e^{-1.086/(n+1)}] + \Delta S^0_{\text{type}}(\text{correction}). \tag{61}$$

Here $\Delta S^0_{\text{type}}(\text{correction})$ is a correction term that makes the calculated entropy from eq. (61) equal to the experimental entropy for a chosen loop type and number of nucleotides (Steger et al., 1984). Thus ΔS^0_{type} (correction) is the experimentally measured ΔS^0 minus the value calculated from eq. (61). For the loop entropies for ten nucleotides given in Table 5.8, these correction terms are: $\Delta S^0_{\text{internal}}(\text{correction}) = 3.48 \text{ kJ mol}^{-1} \text{ K}^{-1}$ $\Delta S^0_{\text{bulge}}(\text{correction}) = 14.28 \text{ kJ mol}^{-1} \text{ K}^{-1}$. and $\Delta S^0_{\text{hairpin}}(\text{correction}) = 8.86 \text{ kJ mol}^{-1} \text{ K}^{-1}$.

The thermodynamics of multibranch loops (junctions) have not been studied systematically yet, but simple models have been proposed to

Table 5.8. Loop-entropy Parameters for Small Loops. Data are from Jaeger et al. (1989). Extra significant figures are given for extrapolation to other loop sizes. The values in parentheses were not measured experimentally, but were extrapolated from the measured data.

Loop size	$\Delta S^0_{\text{internal}}$		$\Delta S^0_{\text{bulge}}$		$\Delta S^0_{\text{hairpin}}$	
	J mol^{-1} K^{-1}	cal mol^{-1} K^{-1}	J mol^{-1} K^{-1}	cal mol^{-1} K^{-1}	J mol^{-1} K^{-1}	cal mol^{-1} K^{-1}
1	—	—	−52.65	−12.57	—	—
2	−55.35	−13.22	−41.85	−10.00	—	—
3	−60.75	−14.51	−47.25	−11.28	−60.75	−14.51
4	−66.15	−15.80	(−56.70)	(−13.54)	−74.25	−17.73
5	(−71.55)	(−17.09)	−64.80	−15.48	−66.15	−15.80
6	(−76.95)	(−18.38)	(−67.50)	(−16.12)	(−68.85)	(−16.44)
7	(−79.65)	(−19.02)	(−70.20)	(−16.77)	−70.20	−16.77
8	(−81.00)	(−19.35)	(−71.55)	(−17.09)	(−74.25)	(−17.73)
9	(−82.35)	(−19.67)	(−72.90)	(−17.41)	(−78.30)	(−18.70)
10	(−85.05)	(−20.31)	(−74.25)	(−17.73)	(−79.65)	(−19.0)

calculate the stability of these loops. A linear approximation for the loop entropy of multihelix junctions is:

$$\Delta S^0 = a + bn + ch \tag{62}$$

with $a = 62.1$ J mol^{-1} K^{-1}, $b = 5.4$ J mol^{-1} K^{-1}, and $c = 1.4$ J mol^{-1} K^{-1}; here n is the number of unpaired bases in the junction, and h is the number of helices (Jaeger et al., 1989). This linear approximation is clearly inappropriate for large loop sizes, and improvements have been reported when a Jacobson–Stockmayer approximation is used instead (Mathews et al., 1997). In the improved model, multibrach loops are calculated as internal loop or bulge loop, depending on whether unpaired bases are found between any two branches (Nussinov et al., 1978), and each additional branch increases the loop size by four unpaired bases to account for the size of the closing base pair (Steger et al., 1984).

8. Prediction of Double-strand Melting Profiles

Prediction of melting profiles for double-stranded RNA or DNA that do not melt in a simple two-state process is possible using the total partition function as described by Poland (1974). From these melting profiles, experimentally accessible data such as temperature dependence of UV absorbance and relative electrophoretic mobility can be determined. Simple models for the loop entropy function are sufficient for prediction of double-strand melting profiles. Only symmetrical internal loops occur, and in most cases, melting starts at the duplex ends. The accuracy of the loop entropy function is therefore not critical; either eq. (60) or eq. (61) can be used.

With most sets of stability parameters available for RNA, DNA, and RNA–DNA hybrids, the prediction of melting transitions in a good agreement with experimental data obtained on duplexes that melt is in the same temperature range as the oligonucleotides studied. The accuracy of prediction is improved for the RNA data set of Turner (1988) if slight corrections (well within the experimental errors of 5%) are made to the stacking entropies (Steger, 1994). Entropy contributions for G–C/G–C stacks are reduced by 2%, entropies for all other stacks are increased by 2%. Extrapolation of the model oligonucleotide data to longer duplex lengths and higher temperatures is more problematic. No satisfactory fit of experimental data determined for the melting of long DNA double strands could be obtained using the parameter set of Breslauer (Steger, 1994).

Prediction of stability for mismatch-containing duplexes has been demonstrated successfully even if, in a first-order approximation, the mismatch is simply regarded as an internal loop. Melting profiles for long duplexes can be predicted to a degree sufficient for most applications. Higher precision may be achieved if the mismatch is treated in a sequence-dependent manner. Werntges et al. (1986) introduced the concept of the "virtual stack" to describe the interaction between the mismatch and its flanking base pairs, and determined stability parameters for mismatches in DNA by fitting experimental melting curves to a partition-function model (Poland and Scheraga, 1970).

9. Prediction of Secondary Structure

The Gibbs free energy is a minimum for a system at equilibrium at constant temperature and pressure. Therefore, accurate calculations of free energies of folded states of nucleic acids can predict the most stable species, and the free energy required to unfold it, or convert it to another conformation. Thermodynamic data to estimate tertiary structures (pseudoknots, junctions, base triples, etc.) are very sparse; thus tertiary structure will not be considered here.

The algorithm commonly used to predict optimal and suboptimal secondary structures of RNA is that of Zuker (1989). It uses the dynamic-programming method (Nussinov et al., 1978) to calculate free energies of secondary structures corresponding to all possible combinations of base pairs. The lowest-free-energy (optimal) secondary structure, plus other low-free-energy (suboptimal) structures are found. Mathews et al. describe the latest results in chap. 6 of this volume.

Whereas the partition function for double-stranded nucleic acids can be easily calculated, the large number of possible secondary structures (including internal loops, hairpin loops, bulges, and junctions) precludes a simple enumeration approach to calculating their partition functions. The commonly used approach to calculate or estimate the partition function, or to calculate favorable secondary structures, is the method of dynamic programming, based on graph theory (Bellman, 1958). In this method, minimum free energies or partial sums of the partition function of a section of the RNA can be calculated recursively from energies or partition functions of smaller sections, reducing the computing effort to the order of N^3 for sequences of length N. The principal requirement for this method is the strict additivity of energies from disjoint sections; this requirement precludes taking into account any tertiary interactions, since these would make free energy inside a section depend on the structure outside it. For similar reasons, the use of dynamic programming is limited to calculating additive thermodynamic quantities such as free energy, or calculating partition functions; no optimization for kinetic data can be achieved in this way.

The Zuker algorithm (Zuker & Stiegler, 1981) has been modified to allow for circular RNA (Steger et al., 1984 and for prediction of suboptimal structures (Steger et al., 1984; Williams and Tinoco, 1986; Zuker, 1989) The most recent version of the Zuker program is available on the internet at [http://www.ibc.wustl.edu/~zuker/rna/form1.cgi]. The web site [http://sun2.science.wayne.edu/~jslsun2/servers/dna/form1.cgi] also includes data for DNA-loop free energies, and permits calculation of DNA secondary structure. An implementation of the Zuker–Nussinov algorithm using a Jacobson–Stockmayer approximation for multibranch loops directly has been reported (Steger et al., 1984); prediction of high-temperature structures is greatly improved by this change. Further changes in the multibranch loop energies result in better agreement between predicted and experimentally determined structures (Mathews et al., 1997).

The ensemble of suboptimal structures generated by these programs can be used for determining base-pair probabilities and calculating UV melting curves, or electrophoretic mobility curves (Schmitz and Steger, 1992). Due to the nature of the free-energy minimization scheme, helices in the structures generally are either fully formed or not present at all; no intermediate states are found. Extension of the energy-minimization approach to obtain the full partition function (McCaskill, 1990) permits calculation of arbitrary thermodynamic data in addition to the base-pair probabilities. For example, specific heat, C_P, could be calculated for comparison with results of differential scanning calorimetry.

Dynamic programming is not capable of finding potential tertiary structure elements, due to the constraint in this type of method that intersecting chords in the representative graph are prohibited. This restriction does not apply to the group of secondary-structure prediction algorithms using combinatorial or Monte Carlo methods. These methods search for the minimum-free-energy structure by finding the most favorable addition of helices or base pairs to partial structures (Abrahams et al., 1990) or by using a random-choice scheme biased by the free-energy decrease for each potential addition (Martinez, 1984). Both methods cannot guarantee actually reaching the minimum-free-energy structure: they can be trapped in high-energy metastable structures. Better convergence to the minimum-free-energy structure is achieved if disruption of helices is permitted, to remove misfolded structures. A high probability of convergence into low-energy structures results from the use of simulated annealing (Schmitz and Steger, 1996).

All of the methods mentioned so far assume that the RNA is in thermal equilibrium and can be adequately described by the minimum-free-energy structure, or the equilibrium partition function. This is not necessarily true, especially for longer RNA, where the kinetics of folding may be slow enough to trap the RNA into some metastable conformation. The biologically relevant folded species may form only as the RNA is being transcribed. Combinatorial and Monte Carlo methods can be modified to optimize for fast folding of an RNA (Mironov et al., 1985; Martinez, 1990), or to simulate the folding during RNA synthesis (Mironov and Kister, 1986). More recent optimization techniques have been applied to the problem of sequential folding by using a modified simulated-annealing approach (Schmitz and Steger, 1996), or a genetic algorithm (Gultyaev et al., 1995) to overcome the problem of high activation barriers for structure refolding at physiological conditions. In the simulated-annealing approach, one needs to average a number of individual foldings for reliable interpretation. Transient structures from the genetic-algorithm approach need to be further analyzed in a separate kinetic model, since the transition probabilities used to generate the structures do not correspond to kinetic rate constants.

A more mathematically rigorous approach to sequential folding (Breton et al., 1997) attempts to describe the probability distribution of structures as evolving according to elementary rate equations in the time interval between each of two transcription events. The set of elementary reactions used for generation of the structure transitions is the same as those employed by Schmitz and Steger (1996), whereas the method relies on a given subset of structures at each chain length approximating the complete set of all possible secondary structures (structure space), instead of sampling the structure space as in the Monte Carlo methods.

Although the accuracy of the available parameter sets (tables 5.3–5.8) is sufficient to give reasonable predictions of RNA secondary structure near 37°C, the accuracy of structure predictions at temperatures significantly higher than 37°C is drastically reduced. Obviously, the approximation that the enthalpy and entropy are independent of temperature is better for a small temperature range. At low temperatures, a folded structure melts to stacked single strands—but, as the temperature increases, the melting is to unstacked strands. The temperature dependence of stacking, and its effect on ΔH and ΔS, thus changes not only Watson–Crick nearest-neighbor parameters but also all the mismatch, terminal-stacking, and loop parameters. In fact, predictions of melting profiles have been found to be less accurate with parameter sets optimized for structure prediction at 37°C (Schmitz and Steger, 1992).

10. Conclusions

We have described the fundamental principles used in obtaining thermodynamic variables of state (Gibbs free energy, enthalpy, entropy) for the interactions of nucleic acids. The main challenge is to relate measured data to the thermodynamic variables. In calorimetry experments, heat effects are measured that directly provide enthalpy changes. It is then necessary to establish the transition or reaction corresponding to the measured enthalpy change. The concentrations and identities of reactants and products must therefore be determined. In equilibrium experiments, the activities of products and reactants at equilibrium are required for determining an equilibrium constant. The equilibrium constant provides the standard Gibbs free energy; the temperature dependence of the equilibrium constant provides the standard enthalpy. Note that equilibrium methods give *standard* thermodynamic variables for defined standard states (where activities are equal to 1). For nucleic acids, activities are defined as equal to concentrations in the limit of infinite dilution.

Characterization of the identities and concentrations of the species in solution is crucial for all the experments. Spectroscopic methods, such as absorption, fluorescence, circular dichroism, nuclear magnetic resonance, and so forth have all been used. As more specific and more accurate analytical methods are applied to nucleic acid solutions, the thermodynamic variables obtained will also become more accurate. When the temperature dependence of each thermodynamic variable is measured, and the standard state conditions are considered, we expect that the thermodynamics variables obtained by different methods will all agree.

Acknowledgments We thank Mr. Luis Comolli, Mr. Ruben Gonzalez, and Dr. Gerhard Steger for carefully reading the manuscript and giving helpful suggestions. Professor Donald Gray, of the University of Texas at Dallas, Professor John SantaLucia of Wayne State University, and Professor Douglas Turner of the University of Rochester kindly provided unpublished materials. The research has been supported by NIH grant GM10840 and DOE grant DE-FG03-86ER60406.

References

Aboul-ela, F., Koh, D., Tinoco, I., Jr., and Martin, F. H. (1985) *Nucleic Acids Res.* **13**, 4811–24.
Abrahams, J. P., van den Berg, M., van Batenburg, E., and Pleij, C. (1990) *Nucleic Acids Res.***18**, 3035–44.
Allawi, H. T., and Santalucia, J., Jr. (1997) *Biochemistry* **36**, 10581–94 (1997).
Antao, V. P., and Tinoco, I., Jr. (1992) *Nucleic Acids Res.* **20**, 819–24.
Antao, V. P., Lai, S. Y., and Tinoco, I., Jr. (1991) *Nucleic Acids Res.* **19**, 5901–5.
Bellman, R. (1958) *Quart. Appl. Math.* **16**, 87–90.
Bloomfield, V. A., Crothers, D. M., and Tinoco, I., Jr. (1974) *Physical Chemistry of Nucleic Acids.* New York: Harper & Row.
Borer, P. N., Dengler, B., Tinoco, I., Jr., and Uhlenbeck, O. C. (1974) *J. Mol. Biol.* **86**, 843–853.
Breslauer, K. J., Frank, R., Blocker, H., and Marky, L. (1986) *Proc. Natl. Acad. Sci. USA* **83**, 3746–50.
Breton, N., Jacob, C., and Daegelen, P. (1997) *J. Biomol. Struct. Dyn.* **14**, 727–40.

Cantor, C. R., and Schimmel, P. R. (1980) *Biophysical Chemistry, Part III. The Behavior of Biological Macromolecules.* San Francisco: W. H. Freeman.
Chaires, J. B. (1997) *Biophysical Chemistry* **64**, 15–23.
Eigen M., and deMayer, L. (1974) In: *Investigations of Rates and Mechanisms of Reactions*, G. G. Hammes (ed.). New York: Wiley-Interscience. Vol. 6, part II, chap. 3.
Felsenfeld, G., and Hirschman, S. Z. (1965) *J. Mol. Biol.* **13**, 407–11.
Frank-Kamentetskii, M. D. (1971) *Biopolymers* **10**, 2623–4.
Goldberg, R. N., and Tewari, Y. B. (1991) *Biophysical Chem.* **40**, 241–61.
Goldstein, R. F., and Benight, A. S. (1992) *Biopolymers* **32**, 1679–93.
Gotoh, O. (1983) *Adv. Biophys.* **16**, 1–52.
Gray, D. M., and Tinoco, I., Jr. (1970) *Biopolymers* **9**, 223–44.
Gray, D. M. (1997) *Biopolymerss* **42**, 783–93.
Gray, D. M. (1997) *Biopolymers* **42**, 795–810.
Gray, D. M., Hung, S.-H., and Johnson, K. H. (1995) In: *Biochemical Spectroscopy*, K. Sauer (ed.). Vol. 246 of *Methods in Enzymology.* San Diego: Academic Press. Pp. 19–34.
Gultyaev, A. P., Batenburg, F. H. D. V., and Pleij, C. W. A. (1995) *J. Mol. Biol.* **250**, 37–51.
He, L., Kierzek, R., SantaLucia, J., Jr., Walter, A. E., and Turner, D. H. (1991) *Biochemistry* **30**, 11124–32.
Henco, K., Sänger, H. L., and Riesner, D. (1979) *Nucleic Acids Res.* **6**, 3041–59.
Inners, L. D., and Felsenfeld, G. (1970) *J. Mol. Biol.* **50**, 373–89.
Jacobson, H., and Stockmayer, W. H. (1950) *J. Chem. Phys.* **18**, 1600–6.
Jaeger, J. A., Turner, D. H., and Zuker, M. (1989) *Proc. Natl. Acad. Sci. USA* **86**, 7706–10.
Klump, H. H. (1990) In: *Landolt-Börnstein VII Biophysics*, W. Sänger (ed.). Berlin: Springer-Verlag. Vol. 1c (Nucleic Acids, Spectroscopic and Kinetic Data), pp. 241–56.
LeCuyer, K. A., and Crothers, D. M. (1994) *Proc. Natl. Acad. Sci. USA* **91**, 3373–7.
Liu, Y., and Sturtevant, J. M. (1997) *Biophysical Chemistry* **64**, 121–6.
Longfellow, C. E., Kierzek, R., and Turner, D. H. (1990) *Biochemistry* **29**, 278–85.
Martinez, H. M. (1984) *Nucleic Acids Res.* **12**, 3223–34.
Martinez, H. M. (1990) In: *Molecular Evolution: Computer Analysis of Protein and Nucleic Acid Sequences*, R. F. Doolittle (ed.). San Diego: Academic Press. Vol. 183, pp. 306–17.
Mathews, D. H., Banerjee, A. R., Luan, D. D., Eickbush, T. H. and Turner, D. H. (1997) *RNA* **3**, 1–16.
McCaskill, J. S. M. (1990) *Biopolymers* **29**, 1105–19.
Mironov, A. A., and Kister, A. E. (1986) *J. Biolmol. Struct. Dyn.* **4**, 1–9.

Mironov, A. A., Dyakonova, L. P., and Kister, A. E. (1985) *J. Biomol. Struct. Dyn.* **2**, 953–62.
Molinaro, M., and Tinoco, I., Jr. (1995) *Nucleic Acids Res.* **23**, 3056–63.
Nussinov, R., Piecznik, G., Grigg, J. R., and Kleitman, D. J. (1978) *SIAM J. Appl. Math.* **35**, 68–82.
Owen, R. J., Hill, L. R., and Lapage, S. P. (1969) *Biopolymers* **7**, 503–16.
Papanicolaou, C., Gouy, M., and Ninio, J. (1984) *Nucleic Acids Res.* **12**, 31–44.
Peritz, A. E., Kierzek, R., Sugimoto, N., and Turner, D. H. (1991) *Biochemistry* **30**, 6428–36.
Petersheim, M., and Turner, D. H. (1983) *Biochemistry* **22**, 256–63.
Poland, D., and Scheraga, H. A. (1970) *Theory of Helix–Coil Transition in Biopolymers*. New York: Academic Press.
Poland, D. (1974) *Biopolymers* **13**, 1859–71.
Pörschke, D., Uhlenbeck, O. C., and Martin, F. H. (1973) *Biopolymers* **12**, 1313–35.
Riesner, D., Maass, G., Thiebe, R., Philippsen, P., and Zachau, H. G. (1973) *Eur J. Biochem* **36**, 76–88.
SantaLucia, J., Jr., Allawi, H. T., and Seneviratne, P. A. (1996) *Biochemistry* **35**, 3555–62.
Schmitz, M., and Steger, G. (1992) *Comp. Appl. Biosci.* **8**, 389–99.
Schmitz, M., and Steger, G. (1996) *J. Mol. Biol.* **255**, 254–66.
Schroeder, S., Kim, J., and Turner, D. H. (1996) *Biochemistry* **35**, 16105–9.
Serra, M. J., Axenson, T. J., and Turner, D. H. (1994) *Biochemistry* **33**, 14289–96.
Serra, M. J., Barnes, T. W., Betschart, K., Gutierrez, M. J., Sprouse, K. J., Riley, C. K., Stewart, L., and Temel, R. E. (1997) *Biochemistry* **34**, 4844–51.
Serra, M. J., Lyttle, M. H., Axenson, T. J., Schadt, C. A., and Turner, D. H. (1993) *Nucleic Acids Res.* **21**, 3845–9.
Steger, G. (1994) *Nucleic Acids Res.* **22**, 2760–8.
Steger, G., Hofman, H., Förtsch, J., Gross, H. J., Randles, J. W., Sänger, H. L., and Riesner, D. (1984) *J. Biomol. Struct. Dyn.* **2**, 543–71.
Steger, G., Müller, H., and Riesner, D. (1980) *Biochim. Biophys. Acta* **606**, 274–84.
Sugimoto, N., Nakano, S., Katoh, M., Matsumura, A., Nakamuta, H., Ohmichi, T., Yoneyama, M., and Sasaki, M. (1995) *Biochemistry* **34**, 11211–6.
Tewari, Y. B., and Goldberg, R. N. (1991) *Biophys. Chem.* **40**, 263–76.
Turner, D. H., Sugimoto, N., and Freier, S. M. (1988) *Annu. Revs. Biophys. Biophys. Chem.* **17**, 167–92.
Varani, G. (1995) *Ann. Rev. Biophys. Biomol. Struct.* **24**, 379–404.
Walter, A. E., and Turner, D. H. (1994) *Biochemistry* **33**, 12715–9.
Walter, A. E., Turner, D. H., Kim, J., Lyttle, M. H., Muller, P., Matthews, D. H., and Zuker, M. (1994) *Proc. Natl. Acad. Sci. USA* **91**, 9218–22.

Werntges, H., Steger, G., Riesner, D., and Fritz, H. J. (1986) *Nucleic Acids Res.* **14**, 3773–90.
Williams, A., Jr., and Tinoco, I., Jr. (1986) *Nucleic Acids Res.* **14**, 299–315.
Wu, M., McDowell, J. A., and Turner, D. H. (1995) *Biochemistry* **34**, 3204–11.
Zuker, M., and Stiegler, P. (1981) *Nucleic Acids Res.* **9**, 133–48.
Zuker, M. (1989) *Science* **244**, 48–52.

6

The Application of Thermodynamics to the Modeling of RNA Secondary Structure

David H. Mathews, Joshua M. Diamond, and Douglas H. Turner

Sequence data are being collected at a rate of over one million nucleotides per day. These data are having a large impact on our understanding of biology and medicine, but the volume of data will require computational analysis to extract the maximum possible information. Consider the scheme shown in fig. 6.1 for determination of structure–function relationships and for development of pharmaceuticals that target RNA. It traces a path from sequence data to inhibitor design. The first step is the gathering of sequence data and basic structural data—from chemical modification and enzymatic cleavage, for example. These data are used to develop a secondary-structure model that is tested by site-directed mutagenesis and chemical-interference studies. The final steps, possibly performed in parallel, are to determine the three-dimensional structure of small structural domains by NMR or crystallography, and to design or select inhibitors.

The determination of secondary structure is therefore an intermediate step between the raw sequence data and the determination of higher-order structure. RNA secondary structure can be predicted in two ways. When a large number of related sequences are known, comparative sequence analysis is the standard technique for determining secondary structure; see the reviews of James et al. (1989) and Pace et al. (1999). When one sequence or few sequences are available, however, thermodynamics can be used to predict secondary structure (Mathews et al., 1999). Thermodynamics can also be used to facilitate the alignment of sequences for comparative analysis (Lück et al., 1996; Mathews et al., 1997).

This chapter will start with a definition of secondary structure and a description of the principles behind thermodynamic prediction of RNA secondary structure. The computer algorithms that predict secondary structure from a single sequence will be described, and some data on the accuracy of folding will be presented. Finally, the use of experimental

178   THERMODYNAMICS IN BIOLOGY

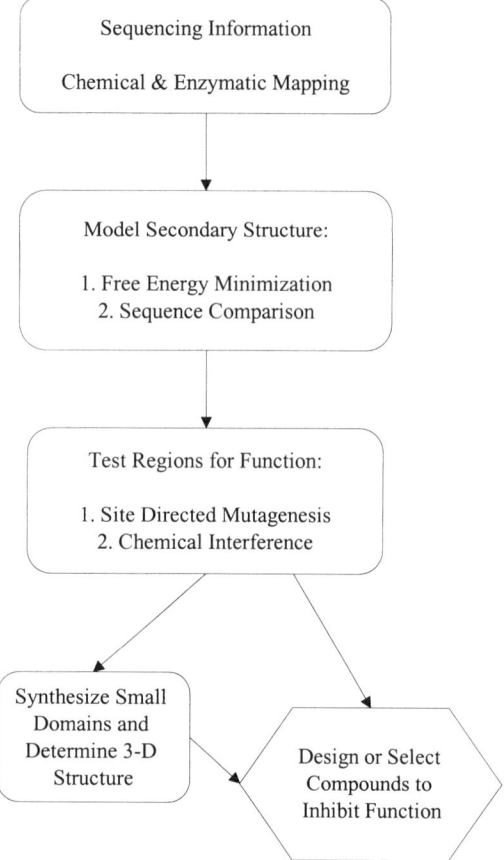

Figure 6.1. A scheme for determination of structure–function relationships and pharmaceutical development. This flowchart shows possible steps for development of a drug from sequence data.

constraints to improve predictions will be discussed, and an example of structure prediction will be presented.

1. Definition of RNA Secondary Structure

For this chapter, RNA secondary structure is defined as the set of Watson–Crick and G–U base pairs in an RNA structure. It can be displayed in two dimensions, as shown in fig. 6.2. The unpaired nucleotides form structural motifs, labeled in fig. 6.2, called internal loops, bulge loops, hairpin loops, and multibranch loops (also called junctions). This definition for secondary structure also includes base pairs involved in pseudoknots

THERMODYNAMICS IN MODELING RNA SECONDARY STRUCTURE 179

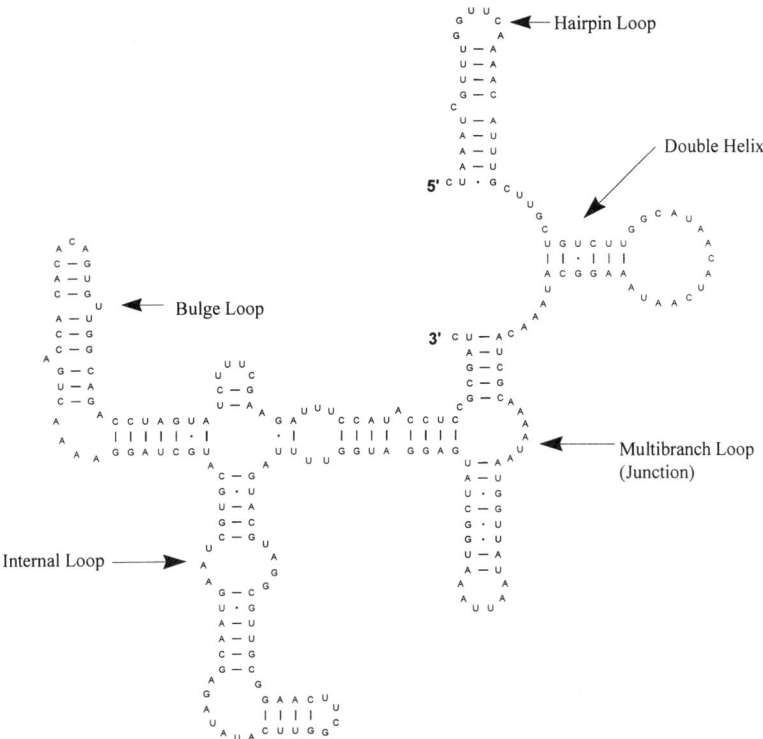

Figure 6.2. The secondary structure model for the 3′ untranslated region of the R2 retrotransposon from *Drosophila melanogaster*. This is a representation of an RNA secondary structure in two dimensions. The base pairs are indicated by lines. The structural motifs are labeled as helixes, internal loops, hairpin loops, bulge loops, and multibranch loops (junctions).

(Pleij, 1994), illustrated in fig. 6.3 as the case of nucleotides in a hairpin-loop base paired with an adjacent region. Including pseudoknots in secondary structure is a controversial choice for several reasons. Base pairs in pseudoknots are not as stable as base pairs in other motifs, and the stability greatly depends on the concentration of Mg^{2+} (Wyatt et al., 1990; Gluick and Draper, 1994). Furthermore, the folding of pseudoknots can be slower than the folding of other base pairs (Zarrinkar and Williamson, 1994; Banerjee and Turner, 1995). These characteristics are more similar to those of tertiary contacts than to base pairs found in other motifs. The pseudoknot can be eliminated, however, by disrupting the pairing in either helix, resulting in either a 5′ or a 3′ hairpin structure with a 3′ or 5′ single-stranded region, respectively. Thus, either hairpin is an element of secondary structure. Choosing which of the two helices is secondary

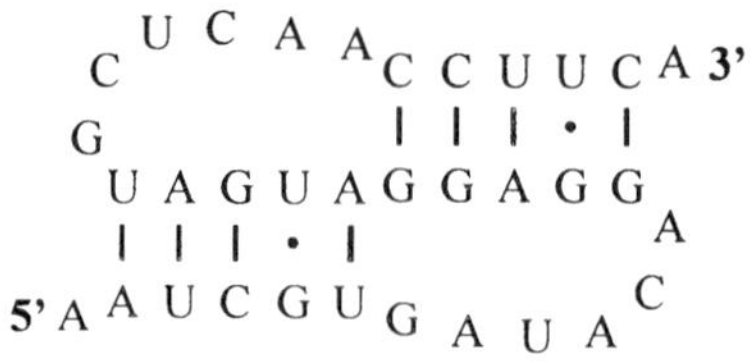

```
         U C A A
       C         C C U U C A 3'
       G         | | | · |
         U A G U A G G A G G
         | | | · |              A
5' A A U C G U G A U A C
```

Figure 6.3. A pseudoknot. This figure illustrates an RNA pseudoknot between hairpin nucleotides and adjacent nucleotides. Pseudoknots can be considered as either secondary structure or tertiary structure.

structure and which is then tertiary structure is not an easy task in all cases. Therefore, it is sometimes more convenient to consider both helices as secondary structural elements.

2. Principles for Thermodynamic Prediction of Secondary Structure

Given enough time, a solution containing RNA strands will reach an equilibrium state. In this state, many conformations may be present, but at different concentrations. For each conformation, there will be an equilibrium with the unpaired state:

$$\text{unpaired state} \rightleftharpoons \text{structure } i. \tag{1}$$

Each conformation i has a Gibbs free-energy difference (ΔG^0) relative to the unpaired state, which quantifies the equilibrium concentrations by the relationship.

$$K_i = e^{-\Delta G_i^0/RT} = \frac{[\text{structure } i]}{[\text{unpaired state}]}, \tag{2}$$

where R is the gas constant (1.987 cal K^{-1} mol^{-1}) and T is the absolute temperature (K). For example, a ΔG_i^0 less than zero indicates that structure i is more favorable than the unpaired state. Thus ΔG^0 quantifies the stability of a structure. Free energy is a function of temperature, so that the relative concentrations of various structures is dependent on temperature.

Equation (2) can be used to derive a relationship between the concentrations of strands in conformation i and another conformation j:

$$\frac{[\text{structure } i,]}{[\text{structure } j,]} = \frac{K_i}{K_j} = e(\Delta G_j^0 - \Delta G_i)/RT. \tag{3}$$

Equation (3) demonstrates that the structure with the lowest free energy (i.e., the most negative) will have the highest concentration.

THERMODYNAMICS IN MODELING RNA SECONDARY STRUCTURE

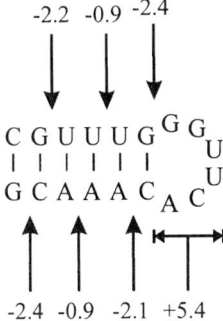

Figure 6.4. Approximating the free energy change of forming a simple secondary structure from an unpaired strand. The free energy of this hairpin is predicted to be −5.5 kcal/mol, based on nearest-neighbor parameters (Mathews et al., 1999). Each free-energy increment is indicated in kcal/mol.

Parameters for Predicting the Free Energy of a Secondary Structure

A simple nearest-neighbor model can be used to predict the standard free-energy change at 37°C for forming a secondary structure; see Xia et al. (1998), Mathews et al. (1999), the review by Xia et al. (1999) and Chapter 5 of this book. In this model, the free-energy increment for forming a base pair depends on the sequence of the adjacent base pair, and the free-energy increments are additive.

Consider the hairpin loop in fig. 6.4. The free-energy change for forming this structure from an unpaired strand is the sum of the free energies of the helical region and the hairpin loop. Using the free-energy increments determined by Xia et al. (1998), the stability of the base paired region will be the sum of five stacking interactions between base pairs:

$$\Delta G^0_{\text{helix}} = \Delta G^0 \begin{bmatrix} C\ G \\ G\ C \end{bmatrix} + \Delta G^0 \begin{bmatrix} G\ U \\ C\ A \end{bmatrix} + 2\Delta G^0 \begin{bmatrix} U\ U \\ A\ A \end{bmatrix} + \Delta G^0 \begin{bmatrix} U\ G \\ A\ C \end{bmatrix}$$

$$= -2.4 - 2.2 - 2 \times 0.9 - 2.1 = -8.5 \quad (\text{kcal/mol}). \quad (4)$$

The hairpin-loop stability is approximated as the sum of an unfavorable term for closing the hairpin loop (initiation), which is a function of loop length, and a favorable term for the stacking of the first mismatch on the last pair (Mathews et al., 1999):

$$\Delta G^0_{\text{hairpin loop}} = \Delta G^0_{\text{initiation}}(6 \text{ nucleotides}) + \Delta G^0_{\text{mismatch}} \begin{bmatrix} G\ G \\ C\ A \end{bmatrix} \quad (5)$$

$$= 5.4 - 2.4 = 3.0 \quad (\text{kcal/mol}).$$

The stability for the entire stem–loop structure is the sum of each substructure.

$$\Delta G^0_{\text{total}} = \Delta G^0_{\text{hairpin loop}} + \Delta G^0_{\text{helix}} = 3.0 - 8.5 = -5.5 \quad (\text{kcal/mol}). \quad (6)$$

The overall free-energy change at 37°C is less than zero, so the equilibrium favors the hairpin structure over the unpaired structure. From eq. (2), the concentration of folded hairpin at 37°C will be almost ten thousandfold higher than that of the unpaired structure.

Importance of the Sequence of Unpaired Regions for Determining the Stability of a Structure

The sequence dependence for the stability of helixes is well determined experimentally (Xia et al., 1998), but the dependence of stability on sequence for unpaired regions is not as well known. Consider the two similar internal loops in fig. 6.5. By changing GA mismatches to AA mismatches, the free-energy increment changes by 4.1 kcal/mol (Wu et al., 1995; Mathews et al., 1999). Thus this sequence change would make the equilibrium constant for folding with this motif less favorable by a factor of about 1000 at 37°C; see eq. (2). Parameters must take this sequence dependence into account to accurately predict stability.

For most motifs, the sequence dependence cannot currently be studied experimentally for each possible sequence. For example, the simple motif of a 2 × 2 internal loop, such as those shown in fig. 6.5, has 4704 possible sequences when closing base pairs are also considered. Three techniques are used to predict the stabilities of those sequences that have not been measured. The first is to study representative sequences, and develop models for the stability of other sequences. This has been used for motifs such as hairpin loops (Serra et al., 1997; Mathews et al., 1999) and 2 × 2 internal loops (Xia et al., 1997; Mathews et al., 1999). Stability can also be based on the occurrence of a sequence in a given motif in a set of known structures. The more frequently the sequence occurs, the higher the stability is assumed to be. This approach was used to assign enhanced stability to certain hairpin loops of four nucleotides (tetraloops) (Jaeger et al., 1989; Walter et al., 1994; Mathews et al., 1999). A third approach is to adjust thermodynamic parameters to improve the accuracy of structures predicted with them. This was used to generate parameters for multibranch loops (Jaeger et al., 1989; Gultyaev et al., 1995; Mathews et al., 1999). The latter two methods have the advantage of simulating extra stability that may be conferred upon motifs by tertiary contacts. For example, tetraloops are known to be involved in tertiary contacts (Michel and Westhof, 1990; Costa and Michel, 1995; Cate et al., 1996; Lehnert et al., 1996).

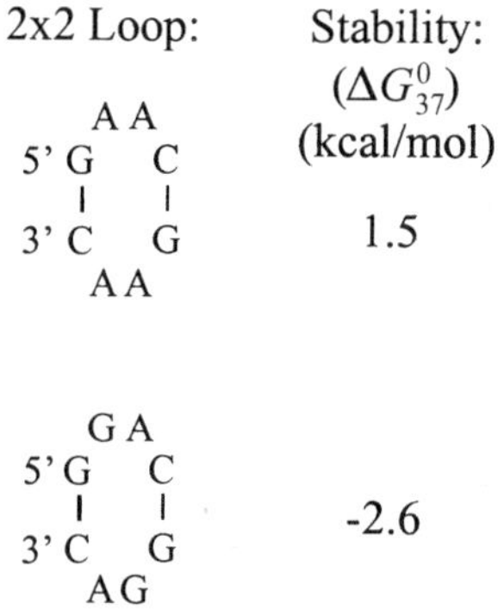

```
2x2 Loop:        Stability:
                  ($\Delta G^0_{37}$)
      A A
5' G     C       (kcal/mol)
    |   |
3' C     G           1.5
      A A


      G A
5' G     C
    |   |            -2.6
3' C     G
      A G
```

Figure 6.5. The stability of two similar 2×2 internal loops. By changing the tandem mismatches from G–A to A–A, the stability of this internal loop changes by 4.1 kcal/mol (Wu et al., 1995).

3. Algorithms for Predicting RNA Secondary Structure

Elements of RNA secondary structure are generally more stable than those of its tertiary structure. This is demonstrated by studies of the denaturation of RNA structures (Banerjee et al., 1993; Jaeger et al., 1993; Laing and Draper, 1994; Crothers et al., 1974; Hilbers et al., 1976; Mathews et al., 1997). Therefore, stability conferred by tertiary structure is mostly not considered by algorithms predicting secondary structure.

One approach for finding the minimum-free-energy RNA structure is to generate every possible secondary structure and then explicitly calculate the free energy of each conformation. With this combinatorial scheme, however, the calculation time of the free energy for all combinations increases exponentially with the number of nucleotides (Zuker, 1989b; Zuker and Sankoff, 1984). It has been estimated that, for a short sequence of 100 nucleotides, there are 3×10^{25} possible structural conformations. With a personal computer running at 500 MHz, it is reasonable to expect to calculate the free energy for about 1000 structures per second; so to calculate the free energy of each structure explicitly would take 10^{15} years (Turner et al., 1988).

Other algorithms have been devised to find the minimum-free-energy structure (Ninio, 1979; Nussinov and Jacobson, 1980; Zuker and Stiegler, 1981; Gultyaev et al., 1995; Rivas and Eddy, 1999). The most popular of these is the dynamic-programming algorithm of Zuker and Stiegler (1981). This algorithm takes advantage of recursion and a simplifying assumption to increase the speed of the calculation to a more reasonable time scale than the combinatorial method.

Consider the nucleotides in a sequence numbered from 5' to 3'. The simplifying assumption used by the dynamic-programming algorithm is that, for nucleotides i, j, k, and l, with $i < j < k < l$, if i is paired with k, then j cannot be paired with l (Zuker and Stiegler, 1981). Similarly, if j were paired with l, then i could not pair with k. This restriction means that the algorithm cannot predict pseudoknots, which are known to occur in natural structures (Pleij, 1994; Gultyaev et al., 1999), such as group I introns (Damberger & Gutell, 1994) and the tmRNA (Williams and Bartel, 1996; Felden et al., 1997). This assumption is reasonable in practice, because most RNAs have only a small fraction of base pairs in pseudoknots. Thus the algorithm correctly predicts a majority of base pairs even for RNA with pseudoknots (Walter et al., 1994; Mathews et al., 1999).

Dynamic Programming to Predict Secondary Structure

There are three steps—fill, traceback, and efn2—in the prediction of structures by the dynamic-programming algorithm. The fill step calculates the value of each position in two arrays: $V_{i,j}$ and $W_{i,j}$, with $i < j$ (Zuker and Stiegler, 1981). The quantity $W_{i,j}$ is the minimal free energy possible for secondary structure of the subsequence from nucleotides i to j. Similarly, $V_{i,j}$ is the minimal free energy possible for the subsequence from nucleotides i to j, but with i and j basepaired. It is assumed that the minimal size of a hairpin loop is three nucleotides. Thus no structure can be formed with less than five nucleotides; so, in practice, the positions of interest are such that $i + 3 < j$. The arrays can be filled by considering every 5-mer, then 6-mer, 7-mer and so forth, until the whole length (1 to N) is filled (Zuker and Sankoff, 1984). The last calculation, of $W_{1,N}$, is the lowest free energy for the entire sequence. Each possible structural motif is checked as the positions are filled, but recursion speeds the calculation. For example, if nucleotide $i + 1$ is paired with $j - 1$, then extension of the helix by a base pair i–j will have a free energy $V_{i,j}$ equal to $V_{i+1,j-1}$, as previously computed, plus the stacking energy of the basepair of i and j.

Once the fill step is complete, the traceback step utilizes this information to determine the base pairs. Essentially, the largest possible regions with $W_{i,j} = V_{i,j}$ (indicating that the lowest free energy for that fragment is equal to the lowest free energy for that fragment with the ends paired) indicate the optimal locations for basepairs between nucleotide pairs i and j. The dynamic-programming algorithm guarantees that the lowest-free-energy structure will be found, given the assumptions inherent in the model.

Because of the necessary approximations, the computed minimal-free-energy structure may not really be the one with the lowest free energy. Therefore, a variation on the dynamic-programming algorithm is used to find representative suboptimal structures that have similar free energy to the lowest-free-energy structure (Zuker, 1989a). This is accomplished with a small change in the fill routine and then tracing back from different starting points for each structure. The generation of suboptimal structures is subject to three parameters: percent sort, maximum number of structures, and window size. Percent sort and maximum number of structures both limit the number of suboptimal structures generated. The algorithm generates suboptimal structures with free energies within percent sort of the lowest free energy up to the maximum number of structures specified. Window size specifies how structurally different each suboptimal structure must be from the others. A window size of zero places no constraint, whereas a larger number will require more variety in the structures. For example, a window size of 3 requires that each suboptimal structure have at least three base pairs separated by three nucleotides from any pair in any other suboptimal structure.

The last step in the current implementation of the algorithm for structure prediction is efn2, which stands for second energy function (Walter et al., 1994). Because the dynamic algorithm is recursive, it is limited in the types of rules it can use in determining free energies. For example, a linear approximation is used for the free-energy dependence on the number of unpaired nucleotides in multibranch loops, but theoretical studies demonstrate a logarithmic dependence (Jacobson and Stockmayer, 1950). Once a limited set of suboptimal structures have been generated by the recursive algorithm, any energy rules can be used to recalculate the free energy of each structure. The structures can then be re-sorted based on these more accurate free energies (fig. 6.6). Typically, one thousand structures are considered in this way.

Accuracy of Dynamic Programming

The accuracy of structure-prediction algorithms is tested by comparing predicted structures to structures determined by comparative sequence analysis. In the most recent comparison (Mathews et al., 1999), secondary structures were predicted for 22 small subunit rRNAs (Gutell, 1994), five large subunit rRNAs (Gutell et al., 1993; Schnare et al., 1996), 309 5S rRNAs (Szymanski et al., 1998), 484 tRNAs (Sprinzl et al., 1998), 91 SRP RNAs (Larsen et al., 1998), 16 RNase P RNAs (Brown, 1998), 25 group I introns (Damberger and Gutell, 1994; Waring and Davies, 1984), and three group II introns (Michel et al., 1989). The ribosomal RNAs and group II introns were split into domains of no more than 700 nucleotides. For this set of structures, 73% of base pairs determined by comparative sequence analysis were correctly predicted in the lowest-free-energy structures. When 750 suboptimal structures were generated for each RNA, one structure contained 86% of the known base pairs, on average. Furthermore, 97% of known base pairs appeared in at least one suboptimal structure.

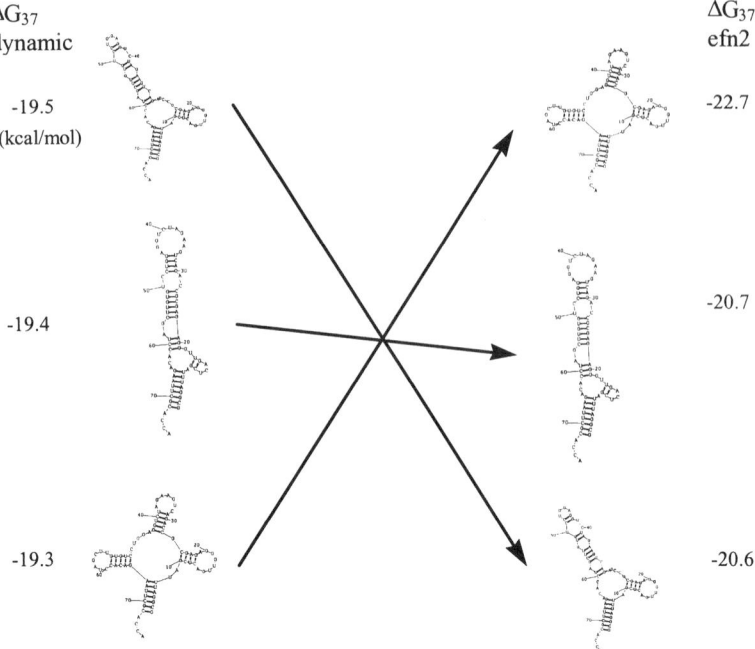

Figure 6.6. Reordering of suboptimal structures based on calculation of ΔG^0 by efn2. This figure illustrates the rearrangement of suboptimal structures according to the free-energy calculation by efn2, which uses different energy rules than the dynamic algorithm. On the left-hand side are three predictions for the secondary structure of a tRNA, one of which has the correct cloverleaf conformation (Sprinzl et al., 1998). The energy rules used by the dynamic algorithm do not predict the cloverleaf as the lowest-free-energy structure, but the efn2 rules predict the correct structure. Note that the most stable structure (with the lowest free energy) is at the top of the stack.

There are many reasons that the algorithm does not predict every base pair. Firstly, the nearest-neighbor model is an approximation, and the stabilities of various motifs are not completely known. As more sequences are studied, these models are expected to improve. Secondly, by using thermodynamics to predict the conformation of an RNA, it is assumed that RNA is in an equilibrium state. It is possible that kinetics are determining some aspects of structure. The extent that kinetics contribute to folding is an area of active interest (Gultyaev et al., 1995; van Batenburg et al., 1995).

The accuracy of predicted structures suggests that thermodynamic structure prediction can be used as a tool in modeling RNA structure. The suboptimal structures can be used to aid in sequence comparison.

Enzymatic cleavage and chemical modification data can be used as constraints to pick out the most accurate of the suboptimal structures.

Use of Constraints in Predicting Secondary Structures

Chemical probes can be used to provide structural information (Burgstaller and Faamulok, 1997)—see also the review of Ehresmann et al. (1987)—as can enzymatic probes, reviewed by Knapp (1989). This information can be used to refine structures determined by either thermodynamic prediction or sequence comparison. For example, RNase VI (cobra venom) cleaves RNA in double-stranded regions without preference to sequence. MB nuclease, RNase T2, and S2 nuclease all cleave single-stranded regions without sequence preference.

The clevage data of RNases can be used directly in the dynamic-programming algorithm to constrain structure prediction. For example, when single-stranded nucleases cleave on both the 5' and 3' sides of a nucleotide, that nucleotide is reliably single-stranded, as demonstrated by many studies (Speek and Lind, 1982; Auron et al., 1982; Kean and Draper, 1985; Tranguch et al., 1994). During structure prediction, all such nucleotides that are between cleavage sites can be entered into the program as single-stranded. The algorithm then forces those nucleotides to be unpaired by assigning large instabilities to any pair involving them. Similarly, nucleotides cleaved on both sides by RNase VI can be forced into the double-stranded state reliably (Speek and Lind, 1982; Tranguch et al., 1994).

For RNA sequences whose secondary structures are poorly predicted without experimental evidence, constraints derived from enzymatic cleavage data can greatly improve the accuracy. In the case of the *E. coli* 5S rRNA (Szymanski et al., 1998), the predicted secondary structure contains only 26% of the base pairs determined by comparative sequence analysis (Mathews et al., 1999). When experimental constraints are used to constrain the prediction, 87% of known base pairs are predicted (Mathews et al., 1999).

Recently, it was reported that flavin derivatives can be used to cleave RNA specifically at uridines involved in G–U base pairs via a photoinduced mechanism (Burgstaller and Famulok, 1997; Burgstaller et al., 1997). These cleavage data can also be used to constrain the prediction of secondary structure (Mathews et al., 1999). For the T4 td group I intron (Jaeger et al., 1993), the accuracy improved from 56% to 83% when expermentally determined FMN cleavages (Burgstaller et al., 1997) were used to constrain the prediction (Mathews et al., 1999).

Chemical-modification reagents, such as 1-cyclohexyl-3-(2-morpholinoethyl)carbodiimide metho-p-toluenesulfonate (CMCT), dimethyl sulfate (DMS), and β-ethoxy-α-ketobutyraldehyde (kethoxal) react with the hydrogen-bonding groups of solvent-exposed bases. CMCT reacts with U and G, DMS with A and C, and kethoxal with G (Ehresmann et al., 1987). Bases accessible to modification occur in single-stranded regions, at the ends of helices, and in or adjacent to G–U pairs. They can be visualized by running end-labeled DNA from reverse transcription on a sequencing gel, because the reverse transcriptase stops at a modified base (Inoue and Cech, 1985; Moazed et al., 1986). Generally, the modifications are quantified as either strong, moderate, or weak. Optimum salt and temperature conditions can be sought for probing so that most tertiary structure has been broken, but secondary structure is still intact (Banerjee et al., 1993; Mathews et al., 1997).

Including chemical-modification data in secondary-structure prediction algorithms is not straightforward. Because single-stranded and certain double-stranded nucleotides can both be accessible to modification by chemical reagents, the structural data cannot be incorporated into the dynamic-programming algorithm. Neither can the data be used to simply pick a single suboptimal structure. The reason for this is illustrated in fig. 6.7. Consider the predicted minimum-free-energy structure shown in part a. Possible suboptimal structures include both structures b and c. In b, domain I has been rearranged in a suboptimal conformation and, in c, domain II is suboptimal. But, because domains I and II are independent, structure d is not generated by the dynamic-programming algorithm. If there were chemical-modification data for this sequence, as indicated by the arrows, structures a, b, and c would all be incompatible with the data. Structure a is incompatible, because the modifications in domains I and II are both in the middle of helices. Structures b and c are incompatible in domains II and I, respectively. The actual structure, d, is fully compatible with the data, but would not be generated as a suboptimal structure.

To incorporate chemical modification data in secondary-structure prediction, the program mix&match can be used (Mathews et al., 1997). First, the dynamic-programming algorithm is used to generate a large number of suboptimal structures (1000 structures with the window size at zero). Then, mix&match selects domains, consistent with chemical-modification data, from suboptimal structures to generate one structure with the lowest free energy that is compatible with the constraints. It is capable of deriving structure 7d from structures 7b and 7c. Mix&match is available from the Turner Lab homepage at http://rna.chem.rochester.edu.

Availability of the Dynamic-programming Algorithm

The World Wide Web is an excellent source for the most up-to-date information about secondary-structure prediction. Versions of the dynamic-programming algorithm are available from two sources (Mathews et al., 1999). A version for personal computers with Windows 95, Windows 98, or Windows NT is available on the Turner

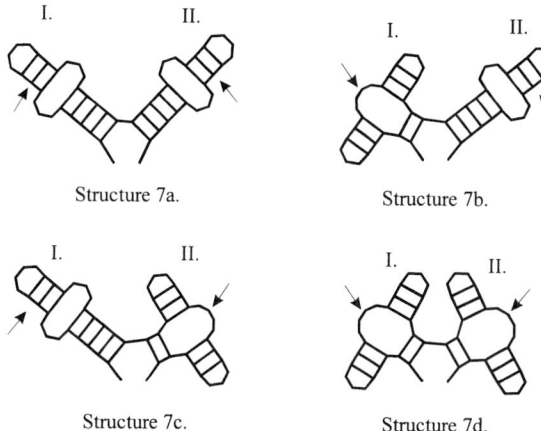

Structure 7a.
Structure 7b.
Structure 7c.
Structure 7d.

Figure 6.7. Suboptimal structures are representative of possible conformations. Four possible secondary-structure conformations are shown for a sequence. Two domains, I and II, exist for each of these structures. Chemical-modification sites are indicated by arrows. Structure 7a is the computed lowest-free-energy structure and 7b and 7c are suboptimal structures. None of these structures is compatible with the chemical-modification data. 7d is an alternative structure that is consistent with the chemical-modification data. It would not be generated by the dynamic-programming algorithm, but can be produced by rearrangement of structures by mix&match (Mathews et al., 1997) with the chemical-modification constraints.

lab homepage at http://rna.chem.rochester.edu (Mathews et al., 1999). Online folding through the World Wide Web version of the algorithm for Unix machines are available at Michael Zuker's homepage, http://www.rpi.edu/~zukerm.

Other Methods for Determining RNA Secondary Structures

A second application of dynamic programming, developed by McCaskill (1990), calculates the partition function for secondary structure. This algorithm can be used to determine the probabilities for the formation of individual base pairs at equilibrium. It makes the same simplifying assumptions about RNA structure as the Zuker algorithm, and therefore also cannot predict psuedoknotted structures.

A dynamic-programming algorithm for secondary-structure prediction capable of predicting pseudoknots has been introduced (Rivas and Eddy, 1999) and a set of thermodynamic parameters for the formation of pseudoknots has been suggested (Gultyaev et al., 1999). The calculation time of the dynamic-programming algorithm by Zuker (1989a) increases on the order of N^3, where N is the number of nucleotides in the sequence. A personal computer can predict the structure of a 1542 nucleotide sequence in about ten minutes (Mathews et al., 1999). The algorithm by Rivas and Eddy (1999) increases in calculation time on the order of N^6, and therefore it is currently limited to predicting structures for sequence shorter than about 100 nucleotides.

An approach using a genetic algorithm that seeks to include kinetic pathways in free-energy minimization has also been developed for RNA secondary-structure determination (Gultyaev et al., 1995; van Batenburg et al., 1995). The genetic algorithm uses the concept of survival of the fittest to predict secondary structure. Structures are generated randomly and then subjected to successive iterations of random change and selection. The most fit structures, those with the lowest free energy, are kept for the subsequent iteration, and less fit structures are discarded.

Kinetic effects are simulated in the genetic algorithm by starting the iterations with nucleotides at the 5' end and then adding 3' nucleotides with successive iterations until all nucleotides are included. Helices are broken in the random-change step with a probability inversely proportional to their stability. Hence, as the sequence elongates (as though it were being transcribed) it may become kinetically trapped in stable local structures. These stable local structures may not be the structures in the minimal-free-energy conformation for the entire molecule, but may be the actual structure in solution for some RNAs.

The genetic algorithm has two drawbacks in comparison with dynamic-programming algorithms. The first is that it is a simulation and therefore may not always reach the same structure if repeated with the same sequence. It is also much slower than most recursive algorithms. Predicting the structure of a 500-nucleotide RNA by the genetic algorithm takes hours (Gultyaev et al., 1995), but the dynamic-programming algorithm takes about two minutes (Mathews et al., 1999).

4. A Case Study: Modeling the Structure of the R2 Retrotransposon 3' Untranslated Region

To illustrate the concepts presented in this chapter, the modeling of the R2 retrotransposon 3' untranslated region (UTR) is presented (Mathews et al., 1997). The R2 element is a non-LTR retrotransposon found at a specific site of the 28S ribosomal gene of most insects (Jakubczak et al., 1991). Retrotransposons are mobile genetic elements that have an RNA stage in their cycle of transposition. They have been divided into two classes: those with long terminal sequence repeats (LTRs) and those without terminal repeats (non-LTR).

THERMODYNAMICS IN MODELING RNA SECONDARY STRUCTURE   191

Retrotransposons are transcribed to RNA, one or more proteins are translated, and then the proteins catalyze the transcription of DNA and the insertion of the DNA transcript into a new site in the genome (Eickbush, 1994; Finnegan, 1997). The protein coded by the R2 element is a reverse transcriptase that also catalyzes the cleavage of one strand of the 28S gene target site. It then utilizes the 3' hydroxyl group of the DNA target to prime reverse transcription in what is termed target-primed reverse transcription (TPRT) (Luan et al., 1993; Luan and Eickbush, 1995).

A necessary condition for the TPRT reaction is recognition of the ~ 250 nucleotides of the 3' UTR by the reverse transcriptase (Luan and Eickbush, 1995). An *in vitro* assay was developed to test the reaction with the *Bombyx mori* (silkworm) R2 retrotransposon (R2Bm) (Luan and Eickbush, 1995). The assay was also used to show that the R2Bm protein can utilize the RNA template of the R2 element in *Drosophila melanogaster* (R2Dm) (Mathews et al., 1997), although the RNAs from the two species have little sequence homology.

It was hypothesized that the structure of the R2 element 3' UTR is recognized by the R2 reverse transcriptase. This would explain the requirement for 250 nucleotides of sequence for TPRT and the recognition by the R2Bm protein of two RNAs with divergent sequences. To start understanding the structure of the R2 3' UTR RNA, a secondary-structure model was developed with the sequences available (ten drosophila species and the *B. mori* sequence). The remainder of this chapter describes the development of this model.

Sequence Comparison

Because of the availability of ten drosophila R2 sequences, sequence comparison was used to model the structure of the 3' UTR. This technique is based on the assumption that, although sequence will drift in evolution, structure will remain constant for RNAs with identical function.

Initially, a sequence alignment was made of the first five available drosophila sequences, shown in fig. 6.8a. Sequences from each species are listed horizontally, and the goal is to have as many nucleotides matching in columns as possible while also minimizing gaps (dashed portions used to keep sequences in phase). Boldfaced regions in fig. 6.8a are those that changed during the next step of the model development when secondary structure was considered. The *B. mori* sequence was too divergent from the drosophila sequences to fit the alignment.

The next step was to find helices maintained in every species. This is possible if the sequence is identical down the alignment, or if there are compensating changes. A compensating change occurs when the sequence in each strand of a helix changes between two species, but still maintains base pairing. For example, if an A–U pair in one species is matched by a G–C pair in another species, or if the orientation

192   THERMODYNAMICS IN BIOLOGY

```
Drosophila melanogaster  (mel)  CUAAAU-CGUUUGGUUCA-AAACAUUUGCUUGCUGUCUUGGCAUAACAUC
          yakuba          (yak)  CUUAAUACGUUUGGUUCACAUACAUCUGCCUGCUGCCUUGGCACAAUAUC
          ananassae       (ana)  CCUAUG-CAC-GGGUUCC-AGAUUAA-GCCUGCUGCCGAAGCAUACCAUC
          pseudoobscura   (obs)  CCUAUA-CAC-AUGUUGGAGAGAAGACGCUUGCUACCUAGGC-UAAUGUG
          takahashii      (tak)  CUGAGG-CGCUUGAUAUAGUGAUUAAUGCCUGC-GUCCUGGCUCAACAUC


mel  A-AUAAAGGCAUAAACAUCGCAAAAUAAUGGUUAUAAUUAAAUGGCUAUGAGGAUGGUUUUAGUACGUAGGCGUUGCGGA
yak  A-A-AAAGGCAUAAACAUCGCACA-UAAUGGUUAUUUA----CGGCUAUGAGGAUGGUUUUAGUACGUAGGCGUUGCGGA
ana  A-AAAUCGGCAUAAAAUUCGCUUAAU--------------------AAAGGAUGGUUUUAGUACGUAGGCGUCCCGGG
obs  A-AAUUAGGUAUAAACAUCGUGGUUGUAAA---------------CUUGAGGUGGGUUUUAGUACGUAUGCGU--GAUU
tak  AAAAUACAGGCAUAAACAUCGCAACUAGC--------------AACAAGGAGGAUGGUUUUAGUACGUAGGCAUUGCGGA


mel  ACU-UC----------GGUUCAUAUAGAGCAAUGAAUCGUGCAUGCUAGGAAAACUGACCACACACAGUGUUGGCAGAC
yak  ACU-UC----------GGUUCGGAUAGAGCAAUGAAUCGUGCAUGCUAGGAA--CUGACCAAA------UAACGCAGCC
ana  ACU-U-----------GUCUCG-------GAUGAAUCGUGCAUGCGGUAUAAUUGGGAUCGAUAACAAAUACCAACUA
obs  ACU-UC----------GUAAUC--------AUGAAUCGUGCAUGCUAGUGGGG-------------UUUGGCCUCCA
tak  ACCCUCAACGUGAAGAAGGUUCAGAUAGAGCAAUGAAUCGUGCAUGCUAGAGUC-------------AUUGGUUCGAC


mel  CUA----------------------------------------GUAUCUUUCGAAGAUUUCCAUACCUCCGCGAUCAAA
yak  CUA----------------------------------------GUAUCUUUCGAAGAUUUCCAUACCUUUGCGAUCAAA
ana  AGUUAUUACUAAAUAUAUCGAAAUACAUAAAUAUCCCGUCCUUACGUACUUU-GAAGAUUUCCAU-CCUCAGCGAACAAA
obs  CUA----------------------------------------GUAUCUUU-GAAGAUUUCCUUCCUCAGCGAUCAAA
tak  CUA----------------------------------------GUAUCUUUCGAAGAUUUCCAUUCCUUCGCGAUCAAA
```

(a)

Figure 6.8. The sequence alignment of R2 RNA in drosophila. Figure 6.8a shows the original sequence alignment made by matching nucleotides. Boldfaced nucleotides are in regions changed for the final alignment based on helices. Figure 6.8b is the final alignment of all ten drosophila species showing helices. Helices are underlined and labeled by letters above so that each strand of a double helix has the same label. Positions of compensating changes are in boldface. (Reprinted from Mathews et al. (1997) with the permission of Cambridge University Press, copyright 1997 by the RNA Society.)

of the pair changes in another species—that is, if $5'$A–U$3'$ changes to $5'$U–A$3'$. At least two compensating changes per helix are necessary to provide reasonable proof based on sequence comparison (James et al., 1989).

Because of the enormous number of possible helices in a 250-nucleotide sequence, free-energy minimization was used to reveal putative basepairs. Each sequence was folded independently and all were compared to find common motifs. Seven helices were discovered by this method, and the alignment was changed to reflect these helices (see fig. 6.8b). In this alignment, the goal is to maintain as many helices as possible vertically down the alignment while still minimizing gaps. Five more drosophila sequences were determined. These sequences confirm the alignment, providing a total of 13 compensating changes (boldfaced in fig. 6.8b). The sequence was also examined for possible pseudoknots, but none were found. An automated system has also been developed to do much of this comparison of minimal-free-energy structures (Lück et al., 1996).

THERMODYNAMICS IN MODELING RNA SECONDARY STRUCTURE 193

```
                                                        A
Drosophila melanogaster  (mel)  CUAAAU-CGUUUGGUUCA-AAACAUUUGCUUGCUGUCUU------GGCAUAAC
           yakuba        (yak)  CUUAAUACGUUUGGUUCACAUACAUCUGCCUGCCUU------GGCACAAU
           ananassae     (ana)  CCUAUG-CAC-GGGUUCC-AGAUUAA-GCCUGCUGCCGA------AGCAUACC
           pseudoobscura (obs)  CCUAUA-CAC-AUGUUGGAGAGAAGACGCUUGCUACCUA------GGC-UAAU
           takahashii    (tak)  CUGAGG-CGCUUGAUAUAGUGAUUAA-------UGCCUGCGUCCUGGCUCAAC
           mauritiana    (mau)  CUAAAA-CGUUUGGUUCA-AAACAUUUGCUUGCUGUCUU------GGCAUAAC
           sechellia     (sec)  CUAAAA-CGUUUGGUUCA-AAACAUUUGCUUGCUGUCUU------GGCAUAAC
           simulans      (sim)  CUAAAA-CGUUUGGUUCA-AAACAUUUGCUUGCUGUCUU------GGCAUAAC
           teissieri     (tei)  CUUAAA-CGUUUGGUUCACAUACAUCUGCCUGCUGCCUU------GGCAUAAU
           ambigua       (amb)  CCGAAACACUAUGUUGGA-AAGAAGACGCUUGCUACCUA------GGCAUAAU


         A           B                                      C           D         E
mel  AUCA-AUAAAGGCAUAAACAUCGCAAAAUAAUGGUUAUAAUUAAAU-GGCUAUGAGGAUGGUUUUAGUACGUAGGCGUUGCG
yak  AUCA-A-AAAGGCAUAAACAUCGCACA-UAAUGGUUAUUUA----C-GGCUAUGAGGAUGGUUUUAGUACGUAGGCGUUGCG
ana  AUCA-AAAUCGGCAUAAAAUUCGCUUAAU--------------------AAAGGAUGGUUUUAGUACGUAGGCGUCCCG
obs  GUGA-AAUUAGGUAUAAACAUCGUGGUUGUAAA----------------CUUGAGGUGGGUUUUAGUACGUAUGCGUGAU-
tak  AUCAAAUACAGGCAUAAACAUCGCAACUAGC---------------AACAAGGAGGAUGGUUUUAGUACGUAGGCAUUGCG
mau  AUCA-AUAAAGGCAUAAACAUCGCAAA-UAAUGGUAAUAUAUAAAU-GGCUAUGAGGAUGGUUUUAGUACGUAGGCGUUGCG
sec  AUCA-AUAAAGGCAUAAACAUCGCAAA-UAAUGGUAAUAUAUUAAAUGGCUAUGAGGAUGGUUUUAGUACGUAGGCGUUGCG
sim  AUCA-AUAAAGGCAUAAACAUCGCAAAACAAUGGUUAUAUAUAAAU-GGCUAUGAGGAUGGUUUUAGUACGUAGGCGUUGCG
tei  AUCA-A-AAAGGCAUAAACAUCGCACAAUAAUGGUUA-AUAC-----GGCUAUGAGGAUGGUUUUAGUACGUAGGCGUUGCG
amb  GUAA-AAUUAGGUAUAAACAUCGCAG--------UUGUAAAC---------UUGAGGU-GG-UUUAGUACGUAGGCGU-GAU


           F             F            E            D                   G
mel  GAACU-UC----------GGUUCAUAUAGAGCAAUGAAUCGUGCAUGC--------UAG-GAAAACUGACCACACACAGUG
yak  GAACU-UC----------GGUUCGGAUAGAGCAAUGAAUCGUGCAUGC--------UAG-GAA--CUGACCAAA------U
ana  GGACU-U-----------GUCUC------GGAUGAAUCGUGCAUGCGGUAUAAUUGG-GAUCGAUAACAAAUACCAACU
obs  -UACU-UC----------GUA--------AUCAUGAAUCGUGCAUGC--------UAG-UGGGG--------------UU
tak  GAACCCUCAACGUGAAGAAGGUUUCAGAUAGAGCAAUGAAUCGUGCAUGC-------UAGAGC---------------AU
mau  GAACU-UC----------GGUUCA-----GCAAUGAAUCGUGCAUGC--------UAG-GAAA-CUGA--------AGUG
sec  GAACU-UC----------GGUUCAGAUAGAGCAAUGAAUCGUGCAUGC--------UAG-GAAA-CUGA--------AGUG
sim  GAACU-UC----------GGUUCAGAUAGAGCAAUGAAUCGUGCAUGC--------UAG-GAAAACUGACCACACGCAG--
tei  GAACU-UC----------GGUUCAGAUAGAGCAAUGAAUCGUGCAUGC--------UAG-GAAA-CUGACCA-----AAUG
amb  GAUGA-CU----------UGUUGAAGUGAAACCAUGAAUCGUGCUCGC--------UAU-UA----------------CG


                                          G                          C           B
mel  UUGGCA-GA---------------------CCUA------GUAUCUUUCGA-AGAUUUCCAUACCUCCGCGAUCAA
yak  AACGCA-GC---------------------CCUA------GUAUCUUUCGA-AGAUUUCCAUACCUUUGCGAUCAA
ana  AAGUUA-UUACUAAUAUAUCGAAAUACAUAAAAUCCCGUCCUUACGUAUCUUU-GA-AGAUUUCCAU-CCUCAGCGAACAA
obs  UGGCCU-CC---------------------ACUA------GUAUCUUU-GA-AGAUUUUCCAUCCUCCGCGAUCAA
tak  UGGUUC-GA---------------------CCUA------GUAUCUUUCGA-AGAUUUCCAUCCUUCGCGAUCAA
mau  UUGACA-GA---------------------CCUA------GUAUCUUUCGAUAGAUUUCCAUACCUCCGCGAUCAA
sec  UUGACA-GA---------------------CCUA------GUAUCUUUCGA-AGAUUUCCAUACCUCCGCGAUCAA
sim  UUGGCA-GA---------------------CCUA------GUAUCUUUCGA-AGAUUUCCAUACCUCCGCGAUCAA
tei  GUGGCA-GC---------------------CCUA------GUAUCUUUCGA-AGAUUUCCAUACCUUUGCGAUCAA
amb  UUGGCCCUU---------------------AAUA------GUAUCUAU-GA-AGAUUUCCCAUCCUCAGCGGUCAA

                                (b)
```

Figure 6.8 *(continued)*

Some regions of sequence are different between species (e.g. the nucleotides 5′ to helix A) and therefore structure for these regions could not be determined by comparison. Base pairs in these regions were found by free-energy minimization. Figure 6.9 is the final structure of the *D. takahashii* 3′ UTR. Lettered helixes are those found in alignment aided by free-energy minimization and the numbered helix was found by free-energy minimization alone. The structure of the related sequence from *D. melanogaster* (R2Dm) is shown in fig. 6.2.

Chemical Modification of D. melanogaster RNA

There were not enough compensating changes in the alignment to prove the *D. melanogaster* structure, so chemical modification was performed to test the model. Data were collected for reaction with CMCT, DMS, and kethoxal at 20°C and 42°C (fig. 6.10). Based on optical melting curves, the

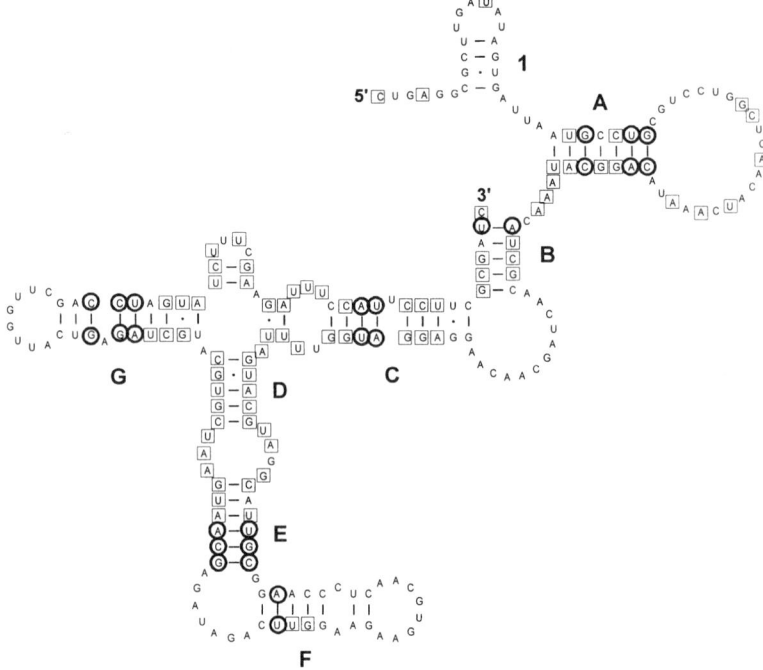

Figure 6.9. The structure of the R2Dt 3′ UTR determined by comparative sequence analysis. This is the secondary structure of the 3′ UTR from the *Drosophila takahashii* R2 determined by sequence comparison and free-energy minimization. Nucleotides invariant throughout the entire alignment are boxed, and positions of compensating changes are circled. Helices determined by sequence comparison are labeled with letters, and the helix determined by free energy alone is numbered. The secondary structure model for the 3′ UTR of R2 from *D. melanogaster* is shown in fig. 6.2.

two reaction temperatures were tested to find conditions where secondary structure is intact, but tertiary structure is not (Mathews et al., 1997). The data support the structure based on sequence comparison and free-energy minimization. At 20°C (fig. 6.10a), all strong modifications occur in single-stranded regions or at the ends of helices. At 42°C (fig. 6.10b), the data are consistent with the opening of helix B. With the exception of the modified C in helix G at 42°C, the strong modifications outside of helix B are consistent with the secondary structure.

Bombyx 3′ UTR Structure

The *B. mori* RNA sequence does not fit the alignment of the drosophila sequences. Therefore another method was required to model its structure.

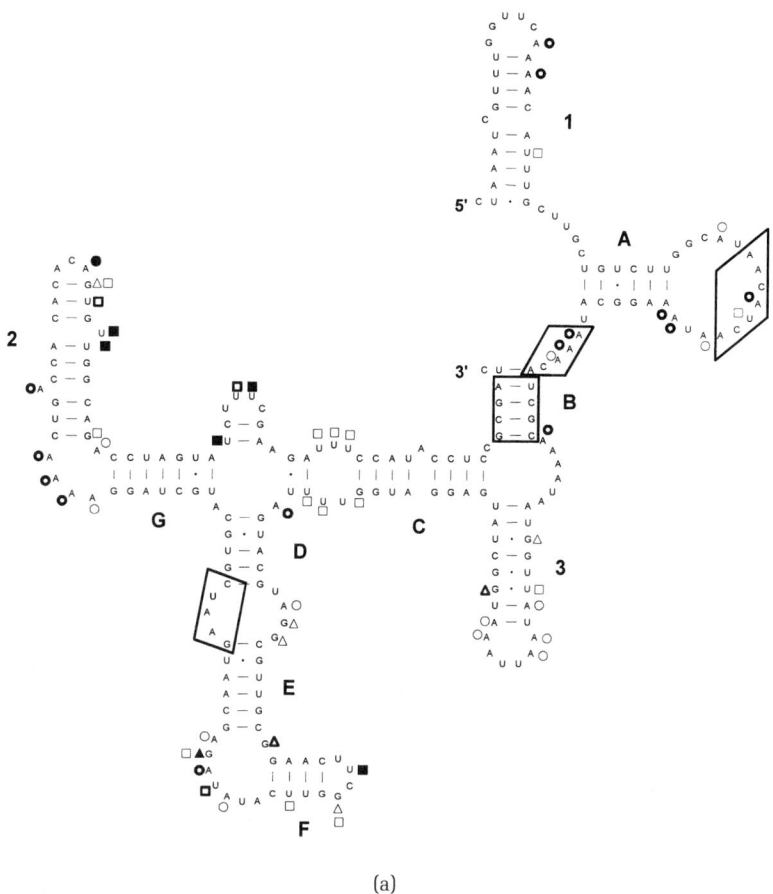

(a)

Figure 6.10. The chemical modification of *D. melanogaster* R2 3′ UTR. Figure 6.10a shows chemical-modification data from 20°C superimposed upon the secondary structure for *D. melanogaster* R2 RNA. Squares represent CMCT modification, circles DMS, and triangles kethoxal. Solid symbols are sites of strong reaction, darkly outlined symbols are moderate, and weakly outlined symbols are weak-modification sites. Boxed regions are those identical in sequence and structure to regions in the *Bombyx* R2 structure (fig. 6.11). Figure 6.10b is the structure with the 42°C modification data superimposed. (Reprinted from Mathews et al., 1997, with the permission of Cambridge University Press, copyright 1997 by the RNA Society.)

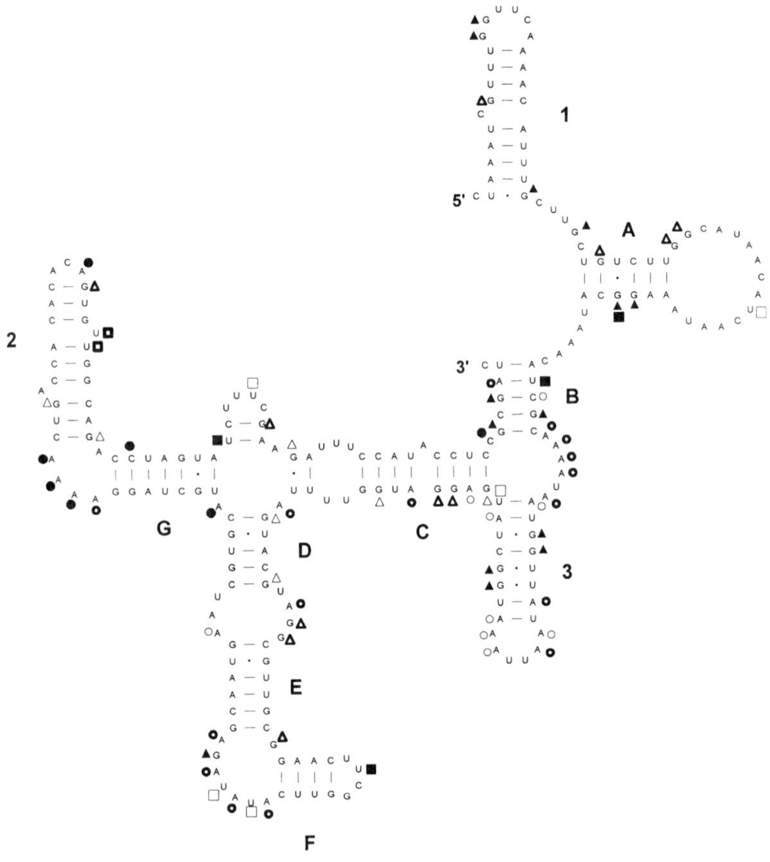

(b)

Figure 6.10 (*continued*)

Chemical-modification data were used to constrain free-energy minimization by the program mix&match (Mathews et al., 1997).

Mix&match requires the input of chemical-modification data and a set of suboptimal structures. Reactivity with CMCT, DMS, and kethoxal were quantified as strong, moderate, or weak with a phosphorimager (Mathews et al., 1997; Zaug & Cech, 1995). Only the strong hits were used in mix&match, because moderate and weak hits can result from alternative structures with a small relative concentration. The dynamic-programming algorithm was used to generate 1000 suboptimal structures with a window size of zero so as to generate as many structural variations as possible.

The lowest-free-energy structure derived from these data is shown in fig. 6.11. There are 11 helices, numbered in the figure, and each strong

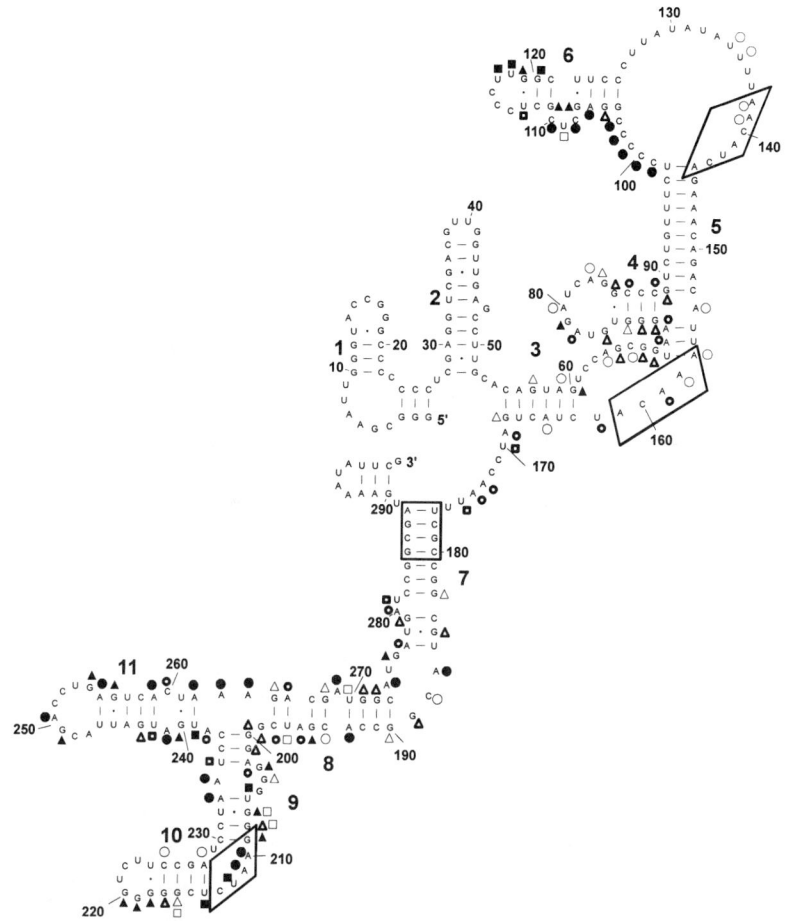

Figure 6.11. The secondary structure of R2Bm 3′ UTR. This figure shows the secondary-structure model derived by constrained free-energy minimization for the 3′ UTR RNA of R2Bm. Chemical modifications are indicated. Squares represent CMCT modification, circles DMS, and triangles kethoxal. Solid symbols are sites of strong reaction, darkly outlined symbols are moderate, and weakly outlined symbols are weak-modification sites. Boxed regions are those identical in sequence and similar in secondary structure to regions in the R2Dm structure (fig. 6.10a). (Reprinted from Mathews et al. (1997) with the permission of Cambridge University Press, copyright 1997 by the RNA Society.)

modification is consistent with the model by design. A majority of moderate and weak modifications are also consistent with the model.

Similarities between R2Bm and R2Dm 3′ UTR RNA

A comparison of the R2Dm, R2Dt, and R2Bm 3′ UTR secondary-structure models reveals four regions of identical sequence and similar structure (boxed in figs. 6.10a and 6.11). These regions may be important for protein recognition of the RNA, consistent with the observation that the R2 *B. mori* protein can recognize and function with *D. melanogaster* RNA. These regions could not be found by sequence comparison alone, but were uncovered by comparing secondary-structure models suggested by free-energy minimization. These models can now serve as a framework for conducting site-directed mutagenesis or deletion experiments to uncover the precise elements required for protein recognition.

5. Summary

Prediction of structure is made possible by the relationship between the Gibbs free-energy change, ΔG^0, and the equilibrium constant (eq. (2)). The free-energy change between a folded and unpaired conformation can be approximated from a set of nearest-neighbor parameters. For helices with Watson–Crick pairs, the parameters were derived by fitting to laboratory measurements of double-helix stability (Xia et al., 1998). For unpaired regions, the parameters were derived by three methods: deducing of rules from studies of model oligonucleotides, basing stability on the frequency of occurrence in a database of known structures, and adjusting stability to optimize the accuracy of structural prediction (Mathews et al., 1999).

Constraints can be used to improve the accuracy of predicted structures. The program mix&match can construct a minimal-free-energy structure consistent with chemical-modification data.

The power of the above method is illustrated by the modeling of RNA secondary structure for the R2 3′ UTR sequence (Mathews et al., 1997). Free-energy minimization aided the modeling of drosophila structures by sequence comparison. Then chemical-modification data were used to constrain free-energy minimization with mix&match to model the structure of *B. mori* R2 RNA. Four regions have identical sequence and similar structure in the drosophila and *B. mori* secondary structures. This supports the hypothesis that the *B. mori* reverse transcriptase is recognizing the structure of the *D. melanogaster* and *B. mori* RNA.

The study of the R2 RNA serves as a model for understanding the structure of a novel RNA. If several related sequences are available, free-energy minimization can be used to facilitate sequence comparison. If no related sequences are available, experimental constraints can be determined to improve the accuracy of a structure found by free-energy minimization.

Secondary-structure prediction is an area of active research. As more measurements of RNA stability are compiled, the free-energy parameters should become known more accurately. The algorithms will also be updated to take advantage of new parameters and advances in technology. Updated information can be found at the Turner Lab Homepage (http://rna.chem.rochester.edu) and Michael Zuker's Homepage (http://www.rpi.edu/~zukerm) on the World Wide Web.

Acknowledgments We thank T. W. Barnes and D. G. Mathews for helpful comments about the chapter.

References

Auron, P. E., Weber, L. D., and Rich, A. (1982) *Biochemistry* **21**, 4700–6.
Banerjee, A. R., Jaeger, J. A., and Turner, D. H. (1993) *Biochemistry* **32**, 153–63.
Banerjee, A. R., and Turner, D. H. (1995) *Biochemistry* **34**, 6504–12.
Brown, J. W. (1998) *Nucl. Acids Res.* **26**, 351–352.
Burgstaller, P., and Famulok, M. (1997) *J. Am. Chem. Soc.* **119**, 1137–8.
Burgstaller, P., Hermann, T., Huber, C., Westhof, E., and Famulok, M. (1997) *Nucl. Acids Res.* **25**, 4018–27.
Cate, J. H., Gooding, A. R., Podell, E., Zhou, K., Golden, B. L., Kundrot, C. E., Cech, T. R., and Doudna, J. A. (1996) *Science* **273**, 1678–85.
Costa, M., and Michel, F. (1995) *EMBO J.* **14**, 1276–85.
Crothers, D. M., Cole, P. E., Hilbers, C. W., and Schulman, R. G. (1974) *J. Mol. Biol.* **87**, 63–88.
Damberger, S. H., and Gutell, R. R. (1994) *Nucl. Acids Res.* **22**, 3508–10.
Ehresmann, C., Baudin, F., Mougel, M., Romby, P., Ebel, J., and Ehresmann, B. (1987) *Nucl. Acids Res.* **15**, 9109–28.
Eickbush, T. H. (1994) In: *The Evolutionary Biology of Viruses*, S. S. Morse, (ed.). New York: Raven Press.
Felden, B., Himeno, H., Muto, A., McCutcheon, J. P., Atkins, J. E., and Gesteland, R. F. (1997) *RNA* **3**, 89–104.
Finnegan, D. J. (1997) *Current Biol.* **7**, R245–8.
Gluick, T. C., and Draper, D. E. (1994) *J. Mol. Biol.* **241**, 246–62.
Gultyaev, A. P., van Batenburg, F. H. D., and Pleij, C. W. A. (1995) *J. Mol. Biol.* **250**, 37–51.
Gultyaev, A. P., van Batenburg, F. H. D., and Pleij, C. W. A. (1999) *RNA* **5**, 609–17.
Gutell, R. R. (1994) *Nucl. Acids Res.* **22**, 3502–3507.

Gutell, R. R., Gray, M. W., and Schnare, M. N. (1993) *Nucl. Acids Res.* **21**, 3055–74.

Hilbers, C. W., Robillard, G. T., Shulman, R. G., Blake, R. D., Webb, P. K., Fresco, R., and Riesner, D. (1976) *Biochemistry* **15**, 1874–82.

Inoue, T., and Cech, T. R. (1985) *Proc. Natl. Acad. Sci. USA* **82**, 648–52.

Jacobson, H., and Stockmayer, W. H. (1950) *J. Chem. Phys.* **18**, 1600–6.

Jaeger, J. A., Turner, D. H., and Zuker, M. (1989) *Proc. Natl. Acad. Sci. USA* **76**, 7706–10.

Jaeger, L., Westhof, E., and Michel, F. (1993) *J. Mol. Biol.* **234**, 331–46.

Jakubczak, J. L., Burke, W. D., and Eickbush, T. H. (1991) *Proc. Natl. Acad. Sci. USA* **88**, 3295–99.

James, B. D., Olsen, G. J., and Pace, N. R. (1989) *Methods Enzymol.* **180**, 227–39.

Kean, J. M., and Draper, D. E. (1985) *Biochemistry* **24**, 5052–61.

Knapp, G. (1989) *Methods Enzymol.* **180**, 192–213.

Laing, L. G., and Draper, J. (1994) *J. Mol. Biol.* **237**, 560–76.

Larsen, N., Samuelsson, T., and Zwieb, C. (1998) *Nucl. Acids Res.* **26**, 177–8.

Lehnert, V., Jaeger, L., Michel, F., and Westhof, E. (1996) *Chem. Biol.* **3**, 993–1009.

Luan, D. D., and Eickbush, T. H. (1995) *Mol. Cell. Biol.* **15**, 3882–91.

Luan, D. D., Kormana, M. H., Jakubczak, J. L., and Eickbush, T. H. (1993) *Cell* **72**, 595–605.

Lück, R., Steger, G., and Riesner, D. (1996) *J. Mol. Biol.* **258**, 813–26.

Mathews, D. H., Banerjee, A. R., Luan, D. D., Eickbush, T. H., and Turner, D. H. (1997) *RNA* **3**, 1–16.

Mathews, D. H., Sabina, J., Zuker, M., and Turner, D. H. (1999) *J. Mol. Biol.* **288**, 911–40.

McCaskill, J. S. (1990) *Biopolymers* **29**, 1105–19.

Michel, F., Umesono, K., and Ozeki, H. (1989) *Gene* **82**, 5–30.

Michel, F., and Westhof, E. (1990) *J. Mol. Biol.* **216**, 585–610.

Moazed, D., Stern, S., and Noller, H. F. (1986) *J. Mol. Biol.* **187**, 399–416.

Ninio, J. (1979) *Biochimie.* **61**, 1133–50.

Nussinov, R., and Jacobson, A. B. (1980) *Proc. Natl. Acad. Sci. USA* **77**, 6309–13.

Pace, N. R., Thomas, B. C., and Woese, C. R. (1999) In: *The RNA World* (2nd ed.) R. F. Gesteland, T. R., Cech, and J. F. Atkins (eds.). Cold Spring Harbor, NY: Cold Spring Harbor Laboratory Press, Pp. 113–41.

Pleij, C. W. A. (1994) *Curr. Opin. Struc. Biol.* **4**, 337–344.

Rivas, E., and Eddy, S. R. (1999) *J. Mol. Biol.* **285**, 2053–68.

Schnare, M. N., Damberger, S. H., Gray, M. W., and Gutell, R. R. (1996) *J. Mol. Biol.* **256**, 701–19.

Serra, M. J., Barnes, T. W., Betschart, K., Gutierrez, M. J., Sprouse, K. J., Riley, C. K., Stewart, L., and Temel, R. E. (1997) *Biochemistry* **36**, 4844–51.

Speek, M., and Lind, A. (1982) *Nucl. Acids Res.* **10**, 947–65.
Sprinzl, M., Horn, C., Brown, M., Ioudovitch, A., and Steinberg, S. (1998) *Nucl. Acids Res.* **26**, 148–53.
Szymanski, M., Specht, T., Barciszewska, M. Z., Barciszewski, J., and Erdmann, V. A. (1998) *Nucl. Acids Res.* **26**, 156–9.
Turner, D. H., Sugimoto, N., and Freier, S. M. (1988) *Ann. Rev. Biophys. Biophys. Chem.* **17**, 167–92.
Tranguch, A. J., Kindelberger, D. W., Rohlman, C. E., Lee, J., and Engelke, D. R. (1994) *Biochemistry* **33**, 1778–87.
van Batenburg, F. H. D., Gultyaev, A. P., and Pleij, C. W. A. (1995) *J. Theor. Biol.* **174**, 269–8.
Walter, A. E., Turner, D. H., Kim, J., Lyttle, M. H., Müller, P., Mathews, D. H., and Zuker, M. (1994) *Proc. Natl. Acad. Sci. USA* **91**, 9218–22.
Waring, R. B., and Davies R. W. (1984) *Gene* **28**, 277–91.
Williams, K. P., and Bartel, D. P. (1996) *RNA* **2**, 1306–10.
Wu, M., McDowell, J. A., and Turner, D. H. (1995) *Biochemistry* **34**, 3204–11.
Wyatt, J. R., Puglisi, J. D., and Tinoco, I., Jr. (1990) *J. Mol. Biol.* **214**, 455–70.
Xia, T., Mathews, D. H., and Turner, D. H. (1999) In: *Prebiotic Chemistry, Molecular Fossils, Nucleosides, and RNA*, D. G. Söll, S. Nishimura, P. B. Moore (eds.). London: Elsevier, Pp. 21–47.
Xia, T., McDowell, J. A., and Turner, D. H. (1997) *Biochemistry* **36**, 12486–97.
Xia, T., SantaLucia, J., Jr., Burkard, M. E., Kierzek, R., Schroeder, S. J., Jiao, X., Cox, C., and Turner, D. H. (1998) *Biochemistry* **37**, 14719–35.
Zarrinkar, P. P., and Williamson, J. R. (1994) *Science* **265**, 918–24.
Zaug, A. J., and Cech, T. R. (1995) *RNA* **1**, 363–74.
Zuker, M. (1989a) *Science.* **244**, 48–52.
Zuker, M. (1989b) In: *Mathematical Methods for DNA Sequences*, M. S. Waterman (ed.). Boca Raton: CRC Press, Pp. 159–84.
Zuker, M., and Sankoff, D. (1984) *Bull. Math. Biol.* **46**, 591–621.
Zuker, M., and Stiegler, P. (1981) *Nucl. Acids Res.* **9**, 133–48.

7

Conformational Fluctuations and Protein Function: The Thermodynamics of a Brownian Motor

Carey K. Bagdassarian and R. Dean Astumian

A protein engaged in its biological function—such as the physical transport of a substrate molecule across an otherwise impermeable cell membrane or the catalysis of a biochemical transformation—can be visualized as traveling along a "reaction" coordinate, with the position on this coordinate defining the three-dimensional conformation of the biopolymer. Indeed, if a membrane-embedded transport protein is to bind substrate on one side of the phospholipid bilayer and carry it to the other, the macromolecule must undergo considerable structural rearrangement. Furthermore, as the protein functions, it moves downhill on a free-energy surface. That is, the driving force for the particular sequence of conformational changes that defines the protein's action is the decrease in free energy associated with these changes. Now, underlying this structural mobility is the fact that the protein is a "breathing" entity capable of conformational fluctuations which are described through the fluctuation spatial positions of all the atoms comprising the protein—these motions arise from thermal noise in the molecule's environment and within the biopolymer itself. We show here how a protein's biological function arises naturally via a coupling between the macromolecular stochastic degrees of freedom—manifested through thermal conformational fluctuations—and the binding and unbinding of the substrate upon which the biopolymer acts. In other words, we investigate the mechanism whereby thermal noise, which in of itself cannot drive a protein *unidirectionally* along a reaction or conformational coordinate, is harnessed or rectified for biological function. Specializing to membrane transport proteins allows for a simplifying mathematical approximation, but a straightforward conceptual extension enables us to consider also enzymatic catalysis of reactions. We will work from a steady-state Fokker–Planck (or more properly, Smoluchowski) equation in the probability density for finding the transporter with a particular conformation and will arrive at the substrate transport velocity as a function of the substrate concentration gradient across a membrane. From this, we develop a thermodynamic force–flux relation for the conformational changes expressed in terms of the protein's dynamical chemical potential when it has some particular conformation. More specifically, if $v(\theta)$ is the velocity of conformational change, and if $\Omega(\theta)$ is the thermodynamic potential when the protein is in conformation θ (we elaborate on the meaning of this choice of a single conformational coordinate below), then the relation is

$$v(\theta) = \frac{d\Omega(\theta)}{d\theta},$$

where all quantities are dimensionless. To the extent that the Fokker–Planck equation (FPE) provides a satisfactory description of protein conformational fluctuations, this expression remains valid for *any* concentration gradient across the membrane, with the form of $\Omega(\theta)$ unchanged. It is this $\Omega(\theta)$, the chemical potential, which will guarantee that the functioning protein—with thermal conformational fluctuations rectified via the coupling to the substrate concentration gradient—flows downhill on a thermodynamic potential surface. Furthermore, and most importantly, we show that no *molecular mechanism* is required to funnel the protein–substrate complex to lower *potential energies* during biological function—the *free-energy* (or chemical-potential) currency for any transformation is provided externally through the substrate gradient. Note the resemblance of the above expression to the force–flux equations of irreversible thermodynamics (DeGroot and Mazur, 1984) and to the phase-field equations used to describe the dynamics of crystal growth from an undercooled fluid (Bagdassarian and Oxtoby, 1994).

To elaborate further, we realize that the proteins under consideration exist, in the simplest cases, in one of two "chemical" states: loaded (or bound) with a substrate molecule or unbound. It is useful to think in terms of a conformation space through which the protein "travels" as its spatial configuration changes; therefore, the protein will be found in one of two conformation spaces, called bound or unbound, depending on the state of the protein. Now, if we constrain our protein to a single space—the unbound one in the absence of substrate, say—then the conformational fluctuations caused by the thermal noise will establish an equilibrium distribution of conformers, with the probability of finding a specific conformation given by a Boltzmann weight in the potential energy of that structure. As such, there is no net flux of conformational probability density in this space. Similarly, if the protein were to have a tremendously large affinity for its substrate, so that it is never released, we would expect a Boltzmann distribution of conformations in the bound space. If we now allow for the natural binding and unbinding of substrate to the biopolymer—with the concomitant "hopping" from one conformation space to the other—we will see that the thermal fluctuations are harnessed to give a directional flow of conformational probability density. This translates into the molecule's macroscopically observable directionality as it functions. In this paper, we consider only passive transport, where the protein carries substrate molecules down the concentration gradient, so that no metabolic energy needs to be expended in the process. We will present a possible mechanism for active transport in a subsequent work.

There has been considerable recent interest in molecular motors, which work in the mesoscopic regime between the microscopic and macroscopic worls; see the reviews by Astumian (1997), Hanggi and Bartusek (1996), and Julicher et al. (1997). The focus of these workers is upon the unidirectional center-of-mass translation of a Brownian particle constrained to move on a one-dimensional railway, and these theories are used, for example, in explaining the motion of the globular protein kinesin on long filamentous microtubules as observed directly by Svoboda et al. (1993). Though our present work explores protein *conformational* fluctuations, and *not* center-of-mass translation of particles having no relevant internal conformational motions, our formalism has a flavor similar to that used in the translocation work, and both processes involve the rectification of thermal noise. Indeed, consider a sphere-like particle on a one-dimensional path constructed from a physical railway, and submerge the entire assembly into a thermal bath, the molecules of which bombard the sphere on a timescale more rapid than the resulting motion of the Brownian particle. Now, the sphere experiences a potential energy of interaction with the underlying rail upon which it moves (more properly, the particle feels a potential of mean force, because the interaction is averaged over the last fluctuations of the thermal bath). One way that unidirectional particle flux can be achieved is through the interconversion or "hopping" between two rail-potential profiles (as in the references above), each composed of a periodically repeating asymmetric sawtooth pattern (see fig.7.2. but introduce asymmetry into each tooth). That is, at any position along the rail, the particle feels one of two possible potentials, and it is this interconversion between the asymmetric patterns that harnesses the unbiased thermal noise acting upon the Brownian sphere. However, there is a "catch": this non-equilibrium switching of potentials requires energy which is provided metabolically through ATP hydrolysis.

In contrast to this, the symmetry needed to drive the transport protein (or enzyme) arises solely from the concentration gradient of substrate across the membrane (or the excess of reactant in a catalyzed reaction). It is this gradient that provides ultimately the free-energy sink to which the protein conformational fluctuations are coupled.

We begin the bulk of this paper with a discussion of the conformational dynamics relevant to the membrane transporter, and follow with a description of these dynamics through the FPE.

1. Protein Conformational Dynamics, Time-correlated Fluctuations, and Membrane Transport

A protein is a biopolymer consisting of 50 to 500 amino-acid subunits linked together in a linear chain whose three-dimensional folded structure or conformation depends on the sequence of the amino-acid building blocks. There are 20 different naturally occurring amino acids, and each is characterized by a unique side group capable of interaction with other amino acids or the protein environment. X-ray crystallographic techniques have been tremendously important in elucidating the structure of proteins; but, because these methods provide an average conformation, the importance of thermal fluctuations around this average has often been neglected. The protein, like any polymer, enjoys fluctuations which take it through a conformation space, and, as stated in the introduction, these fluctuations underscore the conformational mobility crucial for biological function. The timescales and magnitudes of conformational changes vary from atomic motions of 0.001 to 0.5 nm (nanometers) in 10^{-15} to 10^{-1} s, through larger-scale (0.1 to 1 nm) rigid-body motions of groups of amino acids in 10^{-9} to 1 s, to global protein-folding events which may take as long as 10^4 s (McCammon and Harvery, 1987). Molecular-dynamic simulations have been able to capture beautifully the role of shorter-lived fluctuations (Karplus and McCammon, 1986). For example, the protein myoglobin binds oxygen to a heme group deep within its body. If one considered only the x-ray-derived structure, one would find a tremendously high barrier to oxygen diffusion to the protein core—atomic-scale motions in the structural bottleneck are necessary to lower this barrier sufficiently for the myoglobin to function.

Computer simulation also provides us with an important clue to protein function, from which we are led to the idea of time-correlated fluctuations. Consider another protein called pancreatic trypsin inhibitor (Karplus and McCammon, 1986). The rotations of a hexagonal ring from the side chain of a particular deep-lying amino acid can be followed experimentally, and computer work has shown that the rotation requires the following *coordinated* behavior: an initial fluctuation moves a protein segment out of the way of the ring, and the rotation is aided by concerted collisions of the ring with other moieties. These sequential fluctuations can occur because the protein is densely packed with a rich network of steric and longer-ranged intramolecular interactions. In essence, the initial fluctuation reduces the energetic barrier to ring rotation, thereby increasing the probability that rotation takes place. We are *not* implying that the first event *forces* the second through mechanical means: the first only primes energetically the way for the second to occur with greater probability. Indeed, more than 20 years ago, Careri (1974) speculated on the importance of such time-correlated fluctuations and wrote down force–flux relations of irreversible thermodynamics for these processes in terms of free-energy derivatives with respect to relevant conformational variables. This enabled him to study the resulting correlation functions when the fluctuations were small, and he did not need to consider the nature of the free energy. Again, our aim here is to formulate a time-evolution equation for the conformational variable which is driven by the gradient of the conformational chemical potential. The form of this thermodynamic potential will arise from the Fokker–Planck treatment and will be valid when the transporter (or enzyme) is working in a far-from-equilibrium substrate concentration gradient.

If a protein is configured to exhibit these sequential fluctuations, then its motion through conformation space is simplified by being limited to excursions along well-defined paths through the space. A conformational coordinate, θ as introduced above, will indicate the "position" of the protein in the space, and the chemical potential, $\Omega(\theta)$, must be a function of this coordinate. In accord with these ideas, Zwanzig et al. (1992) and Jackson (1993) have investigated the reduction in timescales for protein-folding and conformational changes, respectively, when local fluctuations are labeled as "correct" or "incorrect". That is, the correct structural changes are biased energetically to take the protein along a favored path, thereby making other parts of the conformation space difficult to access. To tie together these ideas, we maintain that a "correct" structural fluctuation will lower the energy barrier to the next correct one, leading the way through conformation space on a path defining the protein's function.

We turn now to the membrane transport protein, where the nature of the conformation path is particularly clear. A living cell is bounded by a membrane which serves as a physical barrier to some essential metabolites and cell wastes. For example, a glucose molecule cannot cross the hydrophobic interior of the membrane. Nonetheless, embedded in this bilayer environment and spanning it is the transport protein which is responsible for carrying substrate from outside the cell to the interior or vice versa (Darnell et al., 1990). As we mentioned before, we focus here on "facilitated" diffusion, where the protein transports substrate down a concentration gradient. The transport protein must exist in two crucial conformations: E_0, which can bind and release substrate from and to the outside of the cell, and E_n, binding and releasing to the interior (the notation will become obvious shortly). In other words, we envision our transporter as a shuttle fluctuating between an outward-facing conformer, E_0, and an inward-facing one called E_n, as drawn in fig. 7.1. It certainly is

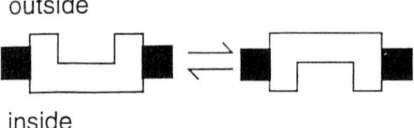

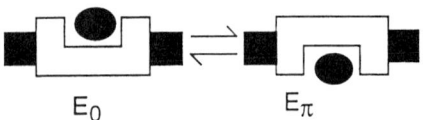

E_0 E_π

Figure 7.1. The outward-facing conformer E_0 (on the left) and the inward-facing E_π (right) are connected through a fluctuation path in conformation space, here summarized by the double arrows. The upper fluctuation sequence is for the unbound space, while the lower is for the bound. The two spaces are linked through the reversible binding and unbinding of substrate.

possible that E_0 and E_n are connected by more than one conformational fluctuation sequence—that is, by more than one path—but we avreage over the probabilistic usage of all roads connecting the two special conformers to arrive at an effective route described by the coordinate θ. Now, the unloaded protein can fluctuate between the two important species by traveling along the conformation path, but it binds substrate only at two singular points on this path: the transporter is "active" when $\theta = 0$ or $\theta = \pi$, defining the outward-facing and inward-facing conformers E_0 and E_n, respectively. In the same way, the loaded transporter fluctuates on the path with release of substrate at $\theta - 0$ or $\theta = \pi$—to the outside or inside of the cell—only (again, the notational choice will become clear below). To sketch how the protein works (when the fluctuations are rectified by the concentration gradient), assume that the substrate density is greater outside so that the transporter binds preferentially a molecule via E_0. Now the protein–substrate complex will undergo *unidirectional* conformational diffusion to E_n and unload into the cell. Finally, the free protein fluctuates back to E_0, allowing the cycle to repeat. If the substrate concentration is greater inside the cell, the sequence of events simply reverses. In the case of the transporter, it is clear that the conformational fluctuations in the bound space serve to move substrate. For an enzyme, the fluctuations would alter the chemical nature of the reactant, and such an extension in principle is easy. However, we are more confident that the binding and unbinding of substrate is limited to well-defined conformers of the transporter.

When the protein is in either the bound or unbound state, we associate with each value of the conformational coordinate θ a potential of mean force. This is the potential energy of the protein in a given conformation averaged over faster-relaxing degrees of freedom such as water and counterion positions. Note that, for a given θ, the only way to distinguish loaded from free protein is through this potential of mean force: the "path length" between E_0 and E_π in both spaces is taken to be the same. Let the potentials associated with the bound and unbound conformation paths be $U_+(\theta)$ and $U_-(\theta)$, respectively, in units of k_BT. Furthermore, take $0 \leq \theta \leq 2\pi$, so that the path is operationally cyclical, which allows for a simpler mathematical treatment. This means that the $\theta = 0$ species E_0 can reach E_π at $\theta = 2\pi$ by clockwise "rotation" from $\theta = 0$ to $\theta = \pi$ or by counterclockwise motion from $\theta = \pi$ (the same conformation as $\theta = 0$) to $\theta = \pi$. However, after solution of the ensuing equations, we will set the potential barrier along one of these legs to infinity, so that the two special conformers will indeed be connected by one path in the sense discussed above. Now, it is reasonable that the $\theta = 0$ and $\theta = \pi$ species, either bound or free, are energetically favored at the level of the potential of mean force, since they are the important forms of the transporter. The potential profiles used are shown in fig. 7.2, and we will see that the results are rather robust to the forms chosen. For the unbound protein, $U(\theta)$ will always have the simple piecewise linear form shown in the figure, but the barrier height c will eventually be set to infinity, closing off conformational diffusion for the free protein betwen $\theta = 0$ and $\theta = \pi$. In conjunction with $U(\theta)$, we investigate three possible forms for $U_+(\theta)$, describing the bound potential. In each case, the barrier b will be set to infinity. The first, shown as the topmost profile in fig. 7.2b, assumes that E_0 and E_π are simply mirror images and have the same potentials of mean force. The middle profile in the figure features a well at $\theta = \pi/2$ but, E_0 and E_π are still energetically equivalent. Finally, in the last profile, $U_+(\pi) > U_+(0)$, and we will show that, if the substrate concentration is greater outside the cell, the bound protein will still travel from $\theta = 0$ to $\theta = \pi$—uphill in *potential energy* (or potential of mean force) but nonetheless downhill in *free energy*. This is what was meant in the introduction by the claim that no mechanistic path is required to channel the protein–substrate complex to lower potential energy. The flow on the thermodynamic free-energy surface is determined via the coupling to the substrate concentration gradient. The solution of the equations will be developed using the $U_-(\theta)$ profile as the model, but results for the bound space follow in a similar fashion. Note the absence (except in the third bound pathway) of tooth asymmetry in our potential profiles—an ingredient crucial to the Brownian motors discussed in the introduction—since the necessary source of symmetry-breaking is through the concentration gradient across the membrane.

210 THERMODYNAMICS IN BIOLOGY

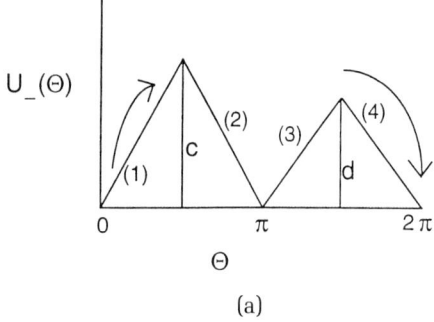

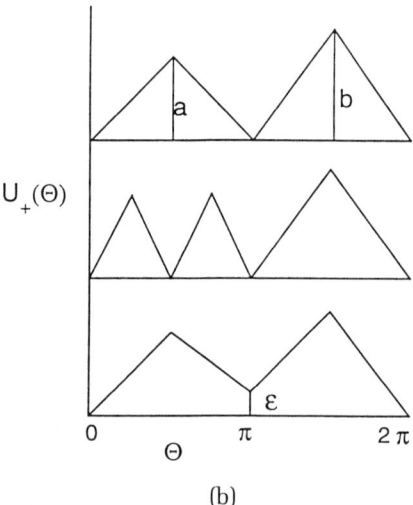

Figure 7.2. Conformational potentials for the unbound (a) and bound (b) spaces.

The dynamics of conformational fluctuations are handled through the Fokker–Planck equation with sink and source terms for the unbinding and binding of substrate. Though the FPE has been used previously to study biopolymer conformational mobility with coupling to other agents (we list in the discussion some examples of previous work where complex processes are described via single coordinates), we postulate here that the *entire* cycle of protein action can be understood through the *reversible* hopping between two conformation spaces, making the fluctuating protein a type of Brownian motor whose *thermodynamics* are easily probed. The implied assumption in such a description is the existence of an equivalent Langevin formulation for protein conformational diffusion in the overdamped limit where the acceleration of the conformational variable is ignored (Serra et al., 1986). This means also that the thermal noise responsible for the diffusion is white.

2. Fokker–Planck (Smoluchowski) Formulation for Conformational Flux

In the simplest case, the FPE relates the time evolution of a probability density to the gradient of the flux in that probability density:

$$\frac{\partial P(x, t)}{\partial t} = -\frac{\partial J(x, t)}{\partial x},$$

where $P(x, t)$ might be, for example, the probability density for finding a Brownian particle at position x at time t. The flux $J(x, t)$ is given as

$$J(x, t) = -D\left(\frac{\partial P(x, t)}{\partial x} + \frac{1}{k_B T \frac{dU(x)}{dx}} P(x, t)\right),$$

with D the diffusion constant for the particle. Note that the flux is driven by a gradient in the probability density and a force $-dU/dx$ acting on the particle ($U(x)$ is the potential energy). For our work, we must consider two coupled FPEs, since we are dealing with a two-state system. Let $P_+(\theta)$ be the probability density that the protein is bound with substrate and has conformation θ; similarly, $P_-(\theta)$ is used to designate the unbound protein. We work in steady state, with $\partial P_+(\theta, t)/\partial t = \partial P_-(\theta, t)/\partial t = 0$, so that the coupled expressions are

$$-\frac{dJ_+(\theta)}{d\theta} - [\gamma_-^0 \delta(\theta) + \gamma_-^\pi \delta(\theta - \pi)] P_+(\theta) + [\gamma_+^0 \delta(\theta) + \gamma_+^\pi \delta(\theta - \pi)] P_-(\theta) = 0,$$

$$-\frac{dJ_-(\theta)}{d\theta} - [\gamma_+^0 \delta(\theta) + \gamma_+^\pi \delta(\theta - \pi)] P_-(\theta) + [\gamma_-^0 \delta(\theta) + \gamma_-^\pi \delta(\theta - \pi)] P_+(\theta) = 0.$$

(1)

Here, $J_+(\theta)$ and $J_-(\theta)$ are the fluxes in conformational probability density when the protein is bound and unbound, respectively. We have divided through by the conformational diffusion constant (which has units of s^{-1} since θ is dimensionless) to render all quantities in (1) dimensionless. The probability density fluxes are given by

$$J_+(\theta) = -\left(\frac{dP_+(\theta)}{d\theta} + \frac{dU_+(\theta)}{d\theta} P_+(\theta)\right), \quad J_-(\theta) = -\left(\frac{dP_-(\theta)}{d\theta} + \frac{dU_-(\theta)}{d\theta} P_-(\theta)\right),$$

(2)

with the bound and unbound potentials of mean force, $U_+(\theta)$ and $U_-(\theta)$, dimensionless because of implicit division by $k_B T$. In the absence of substrate—when the protein is confined to the unbound space—we need solve $-dJ_-(\theta)/d\theta = 0$ to find a Boltzmann distribution of conformational probability density on the fluctuation path: there is no rectification of the noise, and the flux is zero. When substrate is present, the kinetics of the binding and unbinding are described by the delta-function terms which guarantee that these processes occur only at $\theta = 0$ and $\theta = \pi$. The protein accepts substrate with (dimensionless) rates γ_+^0 and γ_+^π at $\theta = 0$ and $\theta = \pi$, respectively; unbinding occurs with rates γ_-^0 and γ_-^π. The unbinding rates are taken to be the same constant at $\theta = 0$ and $\theta = \pi$ but a binding rate is proportional to the volume fraction of substrate: $\gamma_+^\pi \sim \phi_\pi$ for binding from the cell interior, and $\gamma_+^0 \sim \phi_0$ for the exterior, with the same proportionality constant for both. Therefore, $\gamma_+^\pi / \gamma_-^\pi (\gamma_+^0 / \gamma_-^0)$ is an indicator of the relative rate of binding to unbinding at $\theta = \pi$ ($\theta = 0$). Equation (1) needs to be solved for $P_+(\theta)$ and $P_-(\theta)$, and from these the fluxes $J_+(\theta)$ and $J_-(\theta)$ follow.

The piecewise linear potentials $U_+(\theta)$ and $U_-(\theta)$ along with the delta-function form of eq. (1) allow, through a first integration, for the decoupling of the differential equations. Focus first on the unbound potential $U_-(\theta)$ shown in fig. 7.2a, where the piecewise legs have been labeled (1), (2), (3), and (4) and recall that $0 \leq \theta \leq 2\pi$ describes a circle. We will solve for $P_-^{(1)}(\theta)$, $P_-^{(2)}(\theta)$, $P_-^{(3)}(\theta)$, and $P_-^{(4)}(\theta)$, representing the probability distributions on the first, second, third, and fourth legs of the unbound potential, respectively. To obtain $P_-^{(1)}(\theta)$, integrate (with θ the variable of integration) the second expression in eq. (1) from $3\pi/2$ on segment (4) to 2π on (4), and then from 0 on segment (1) to an arbitrary θ on (1). This first integration spans the delta function at $\theta = 0$ ($\equiv 2\pi$) but not the one at $\theta = \pi$, and this path is shown by the arrows in fig. 7.2a. We are left with

$$-J_-^{(1)}(\theta) + J_-^{(1)}(0) - J_-^{(4)}(2\pi) + J_-^{(4)}(3\pi/2) - \gamma_+^0 P_-(0) + \gamma_-^0 P_+(0) = 0. \quad (3)$$

Here $J_-^{(i)}(\theta)$ refers to the flux on the ith segment as defined on the piecewise potential. Note that, through the delta function, the integration has introduced the densities $P_-(0)$ and $P_+(0)$. Though formally "one-half" of the delta function lies in segment (4) and the other half in segment (1), we have used the fact that the probability density on the fourth segment at $\theta = 2\pi$ must equal the density on segment (1) at $\theta = 0$. Now, $J_-^{(1)}(0)$, $J_-^{(4)}(2\pi)$, and $J_-^{(4)}(3\pi/2)$ are constants, and

$$-J_-^{(1)}(\theta) = \frac{dP_-^{(1)}(\theta)}{d\theta} + \frac{2c}{\pi} P_-^{(1)}(\theta), \quad (4)$$

Since $dU_-/d\theta = 2c/\pi$ in this leg; clearly, the differential equation (3) for $P_-^{(1)}(\theta)$ is decoupled from that for $P_+(\theta)$. In a similar manner, we obtain the equation for $P_-^{(2)}(\theta)$ by continuing the integration above through $\pi/2$ to θ somewhere in segment (2). The result is, with anticipation that $J_-^{(1)}(\pi/2) = J_-^{(2)}(\pi/2)$ (since $\theta = \pi/2$ does not serve as a point for substrate binding or unbinding, and we want no singularity in the flux here):

$$\frac{dP_-^{(2)}(\theta)}{d\theta} - \frac{2c}{\pi} P_-^{(2)}(\theta) + A = 0, \quad (5)$$

where $A \equiv J_-^{(1)}(0) - J_-^{(4)}(2\pi) + J_-^{(4)}(3\pi/2) - \gamma_+^0 P_-(0) + \gamma_-^0 P_+(0)$. If we begin the integration at $\pi/2$ on segment (2) and span the delta function at $\theta = \pi$, we find the defining equations for $P_-^{(3)}(\theta)$ and $P_-^{(4)}(\theta)$:

$$\frac{dP_-^{(3)}(\theta)}{d\theta} + \frac{2d}{\pi} P_-^{(3)}(\theta) + B = 0, \tag{6}$$

$$\frac{dP_-^{(4)}(\theta)}{d\theta} - \frac{2d}{\pi} P_-^{(4)}(\theta) + B = 0, \tag{7}$$

with the understanding that $J_-^{(3)}(3\pi/2) = J_-^{(4)}(3\pi/2)$ and

$$B \equiv J_-^{(2)}(\pi/2) - J_-^{(2)}(\pi) + J_-^{(3)}(\pi) - \gamma_+^\pi P_-(\pi) + \gamma_-^\pi P_+(\pi).$$

The four differential equations, which are in short

$$J_-^{(1)}(\theta) = A, \qquad J_-^{(2)}(\theta) = A, \qquad J_-^{(3)}(\theta) = B, \qquad J_-^{(4)}(\theta) = B,$$

are trivially integrated to give

$$P_-^{(1)}(\theta) = \frac{\pi}{2c} A + \left(P_-(0) + \frac{\pi}{2c} A\right) e^{-2c\theta/\pi}, \tag{8}$$

$$P_-^{(2)}(\theta) = \frac{\pi}{2c} A + \left(P_-(\pi) - \frac{\pi}{2c} A\right) e^{2c(\theta-\pi)/\pi}, \tag{9}$$

$$P_-^{(3)}(\theta) = -\frac{\pi}{2d} B + \left(P_-(\pi) + \frac{\pi}{2d} B\right) e^{-2d(\theta-\pi)/\pi}, \tag{10}$$

$$P_-^{(4)}(\theta) = \frac{\pi}{2d} B + \left(P_-(0) - \frac{\pi}{2d} B\right) e^{2d(\theta-2\pi)/\pi}, \tag{11}$$

where we have demanded that $P_-^{(1)}(0) = P_-^{(4)}(2\pi) \equiv P_-(0)$ and $P_-^{(2)}(\pi) = P_-^{(3)}(\pi) \equiv P_-(\pi)$, allowing for evaluation of the constants of integration. Finally, setting $P_-^{(1)}(\pi/2) = P_-^{(2)}(\pi/2)$ and $P_-^{(3)}(3\pi/2) = P_-^{(4)}(3\pi/2)$ determines A and B, respectively, so that the fluxes are

$$J_-^{(1)} = J_-^{(2)} = \frac{c}{\pi} \frac{e^{-c}}{1-e^{-c}} [P_-(0) - P_-(\pi)], \tag{12}$$

$$J_-^{(3)} = J_-^{(4)} = \frac{d}{\pi} \frac{e^{-d}}{1-e^{-d}} [P_-(\pi) - P_-(0)]. \tag{13}$$

Note that the flux through segments (1) and (2) is constant and established by the difference in probability densities at $\theta = 0$ and $\theta = \pi$—where the protein can interact with the substrate—for the unbound biopolymer. With eqs. (12) and (13) and the definitions of A and B given above, we learn two additional relations which will be needed to find the values of $P_-(0), P_-(\pi), P_+(0)$, and $P_+(\pi)$:

$$\gamma_+^0 P_-(0) = \gamma_-^0 P_+(0), \tag{14}$$

$$\gamma_+^\pi P_-(\pi) = \gamma_-^\pi P_+(\pi). \tag{15}$$

It is important to realize that the conformational probability-density flux for the unbound protein is determined solely via probabilities on the unbound fluctuation path—in this sense, the flux in the unbound space is decoupled from that in the bound. Nonetheless, the quantities $P_-(0)$, $P_-(\pi)$, $P_+(0)$, and $P_+(\pi)$ must be evaluated from a consideration of the two conformation spaces together. Note also that the flux from $0 \leq \theta \leq \pi$ has a sign opposite to that from $\pi \leq \theta \leq 2\pi$. This is understood easily. Say $P_-(0) > P_-(\pi)$, so that $\theta = 0$ serves as a source of probability density. The flow has two possible paths, as discussed above: from $\theta = 0$ to $\theta = \pi$ with a positive sign for the flux, or from $\theta = 2\pi (\equiv 0)$ to $\theta = \pi$ with a negative sign. Both these fluxes "empty" into the sink at $\theta = \pi$.

Again, as we discussed above, we will shut off the flux in the unbound space for $0 \leq \theta \leq \pi$ by letting the barrier height c tend to ∞. This eliminates the need for equation (12). How do we determine $P_+(0)$ and $P_+(\pi)$? We must solve the first expression in eq. (1) using a profile for $U_+(\theta)$. The solution for the *top-most* potential shown in fig. 7.2b is readily obtained from the results for the unbound protein: simply interchange "$-$" and "$+$" labels and replace the barrier heights "c" with "a" and "d" with "b". In order to have only one fluctuation path through the bound space as well, let $b \to \infty$. Finally, as a simplification that allows for more compact expressions, we set the barrier heights $a = d \equiv a$. Since both the probability densities and fluxes disappear for $0 \leq \theta \leq \pi$ in the case of the unbound protein and for $\pi \leq \theta \leq 2\pi$ for the bound, in this simplified scheme we are led to

$$P_+^{(1)}(\theta) = -\frac{\pi}{2a}J_+ + \left(P_+(0) + \frac{\pi}{2a}J_+\right)e^{-2a\theta/\pi}, \qquad (16)$$

$$P_+^{(2)}(\theta) = \frac{\pi}{2a}J_+ + \left(P_+(\pi) - \frac{\pi}{2a}J_+\right)e^{2a(\theta - \pi a)/\pi}, \qquad (17)$$

$$P_-^{(3)}(\theta) = -\frac{\pi}{2a}J_- + \left(P_-(\pi) + +\frac{\pi}{2a}J_-\right)e^{-2a(\theta - \pi)/\pi}, \qquad (18)$$

$$P_-^{(4)}(\theta) = \frac{\pi}{2a}J_- + \left(P_-(0) - \frac{\pi}{2a}J_-\right)e^{-2a(\theta - 2\pi)/\pi}, \qquad (19)$$

where

$$J_+ = \frac{a}{\pi}\frac{e^{-a}}{1 - e^{-a}}[P_+(0) - P_+(\pi)], \qquad (20)$$

from $0 \leq \theta \leq \pi$ on the bound fluctuation path, and

$$J_- = \frac{a}{\pi}\frac{e^{-a}}{1 - e^{-a}}[P_-(\pi) - P_-(0)], \qquad (21)$$

for $\pi \leq \theta \leq 2\pi$ when the protein is unloaded. It is clear that $J_+ \equiv J_+^{(1)} = J_+^{(2)}$, and $J_- \equiv J_-^{(3)} = J_-^{(4)}$.

Now, we have two further requirements which must be met, and with eqs. (14) and (15), these conditions establish the four remaining quantities $P_-(0)$, $P_-(\pi)$, $P_+(0)$, and $P_+(\pi)$. First, we demand that all flux leaving one space appears in the other. That is, if the protein conformation is diffusing from $\theta = 0$ to $\theta = \pi$ when the biopolymer is loaded, the flux into the sink at $\theta = \pi$ must appear at $\theta = \pi$ in the unbound space: this is expressed as

$J_- = J_+$. Secondly, the probability densities must be normalized over both spaces:

$$\int_0^\pi P_+(\theta)d\theta + \int_\pi^{2\pi} P_-(\theta)d\theta = 1,$$

where $P_+(\theta)(P_-(\theta))$ is short for the two density legs in the bound (unbound) space. We get, when the barrier heights to diffusion in both states are the same,

$$P_-(\pi) = \frac{1}{\pi}\frac{a}{1-e^{-a}}\frac{1}{1+\gamma_+^\pi/\gamma_-^\pi}, \qquad (22)$$

$$P_-(0) = \frac{1}{\pi}\frac{a}{1-e^{-a}}\frac{1}{1+\gamma_+^0/\gamma_-^0}, \qquad (23)$$

$$P_-(\pi) = \frac{1}{\pi}\frac{a}{1-e^{-a}}\frac{\gamma_+^\pi/\gamma_-^\pi}{1+\gamma_+^\pi/\gamma_-^\pi}, \qquad (24)$$

$$P_+(0) = \frac{1}{\pi}\frac{a}{1-e^{-a}}\frac{\gamma_+^0/\gamma_-^0}{1+\gamma_+^0/\gamma_-^0}. \qquad (25)$$

Note that the substrate concentration is introduced through γ_+^0 and γ_+^π. This quantifies the scheme we summarized above for the functioning protein. For example, say $\gamma_+^0 > \gamma_+^\pi$, because the substrate concentration is larger outside the cell (recall that $\theta = 0$ refers to the cell's exterior). Then $P_-(\pi) > P_-(0)$ and $J_- > 0$: the free protein is diffusing from $\theta = \pi$ to $\theta = 2\pi$. That is, the unloaded transporter undergoes net unidirectional conformational diffusion from the inward-facing conformer to the outward-facing one. From eqs. (24) and (25), $P_+(0) > P_+(\pi)$, so that $J_+ = J_- > 0$: $\theta = 0$ serves as the source of bound conformer which diffuses from the outward-facing species to the inward-facing one.

One final point needs to be made. The transporter effectively shuts off when $J_+ = J_- = 0$, and this is established when $P_+(0) = P_+(\pi)$ and $P(0) = P_-(\pi)$. (This simply means that the probabilities of diffusion from $\theta = 0$ to $\theta = \pi$ and from $\theta = \pi$ to $\theta = 0$ are the same for the loaded protein, say.) From eqs. (14) and (15), we see that $\gamma_+^0\gamma_-^\pi/\gamma_+^\pi\gamma_-^0 = 1$. Recall that γ_+^0 and γ_+^π are proportional to the substrate concentrations outside and inside the cell, respectively, with the same proportionality constant. Also, the substrate unbinding rates at $\theta = 0$ and $\theta = \pi$ are equal, so that $\gamma_-^0 = \gamma_-^\pi$. For this symmetric transporter, then, there is no flux when $\phi_0 = \phi_\pi$—when the outside and inside substrate concentrations are the same.

This completes the solution of the coupled differential equations in (1) when the top-most potential profile in fig. 7.2b is used. Now, as will be shown below, the flux of conformational probability density is proportional to the transport rate. For $\gamma_+^0/\gamma_-^0 = 10\phi_0$ (top curve) and $\gamma_+^0/\gamma_-^0 = \phi_0$ (middle curve), fig. 7.3 shows J_+, or the transport rate, as a function of the external substrate concentration ϕ_0 when the inside of the cell is sub-

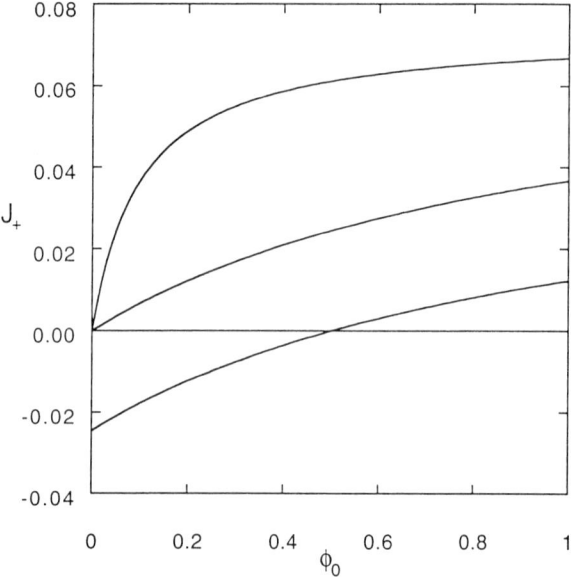

Figure 7.3. Flux as a function of external substrate concentration. For the top two curves, the cell interior is substrate-free; for the bottom curve, $\phi_\pi = 0.5$. See text for details.

strate-free ($\phi_\pi = 0$). We have taken the barrier $a = 2$. It is clear that, as the intrinsic binding rate (the proportionality constant in $\gamma_+^0 \sim \phi_0$) increases with respect to that of unbinding, the transport rate saturates more readily to maximal velocity. This is because $P_+(0) - P_+(\pi)$ becomes maximized, meaning that we are saturating the probability that the transporter is loaded at $\theta = 0$. When the inside of the cell does contain substrate ($\phi_\pi = 0.5$; see the bottom curve of fig. 7.3), the transport is first from the interior to the exterior—corresponding to a negative flux—until it reverses for $\phi_0 > 0.5$. For this last curve, $\gamma_+^0/\gamma_-^0 = \phi_0$ and $\gamma_+^\pi/\gamma_-^\pi = \phi_\pi$.

To solve eq. (1) using the middle potential $U_+(\theta)$ shown in fig. 7.2b—the one with the well at $\theta = \pi/2$—and, again, $U_-(\theta)$ from fig. 7.2a, we proceed in exactly the same way as above except that there will now be four density legs in the bound space for $0 \leq \theta \leq \pi$. These need not be reproduced here. Nonetheless, when the barrier heights b and c are set to infinity in $U_+(\theta)$ and $U_-(\theta)$, respectively, and $a = d$, we learn that the fluxes for $0 \leq \theta \leq \pi$ in the bound space and for $\pi \leq \theta \leq 2\pi$ in the unbound are exactly the same as for the previous potential. That is, we recover again, when there is a well in $U_+(\theta)$, eqs. (20) and (21) for the fluxes. Indeed, $P_-(\pi)$, $P_-(0)$, $P_+(\pi)$, and $P_+(0)$ here are also given by eqs. (22)–(25), and so fig. 7.3 summarizes the flux behavior for this bound potential as well. It is important to realize that such a potential may not correspond to physical reality: it serves only as an illustration, and the first potential is certainly more realistic, since there is no reason to populate the $\theta = \pi/2$ conformer when the transporter is in equilibrium.

We are beginning to see that the results obtained are robust to the forms chosen for $U_+(\theta)$ when $U_+(0) = U_+(\pi)$. This is because the conformational diffusion from $\theta = 0$ to $\theta = \pi$ when the protein is loaded and in the bound space—where the various trial potentials are introduced—is governed by the probability densities at these two points on the bound fluctuation path only, and this is the decoupling between the two spaces. In essence, if we know $P_+(0)$ and $P_+(\pi)$, all we need to do is solve the FPE for conformational diffusion between $\theta = 0$ and $\theta = \pi$ using these densities at the endpoints as boundary conditions. The flux turns out to be sensitive to the barrier height but not to the details of the profile. Indeed, use of an asymmetric tooth between $0 \le \theta \le \pi$ would leave the form of the flux unchanged (though the values of $P_+(0)$ and $P_+(\pi$ do change). Notice that the flux expressions include the multiplicative prefactor e^{-a}. Such an exponential in the barrier height also appears in traditional kinetic treatments of the transport (or catalysis) problem, because these schemes implicitly designate a special conformer of the protein as the active form. That is, kinetic analysis assumes, as we do in our stochastic development, that substrate binds and unbinds to particular conformers of the transporter, and we can match (in the limit of large barrier heights) our results thus far to those of enzyme kinetics or rate theories. Also the Michaelis–Menten-like (Darnell et al., 1990) saturation of the transport rate with increasing substrate concentration arises in both treatments.

The third and final potential profile shown in fig. 7.2b is interesting, because $U_+(\pi)$ exceeds $U_+(0)$ by the quantity ε, and if, the transporter is diffusing from $\theta = 0$ to $\theta = \pi$, the substrate concentration gradient is used to drive the protein–substrate complex to a higher potential of mean force. We stress that, nonetheless, the biopolymer complex travels downhill in *free energy* when diffusing from $\theta = 0$ to $\theta = \pi$. It is not surprising that, for a given ϕ_π inside of the cell, a larger ϕ_0, as compared with the case where $U_+(0) = U_+(\pi)$, is required for net transport to the interior. This is because the barrier to diffusion from $\theta = \pi$ to $\theta = 0$ is less than that from $\theta = 0$ and $\theta = \pi$. Solution of this problem yields $P_+^{(1)}$, $P_-^{(3)}(\theta)$, and $P_0^{(4)}(\theta)$ identical to those given by eqs. (16), (18), and (19), respectively, but

$$P_+^{(2)}(\theta) = \frac{\pi}{2(a-\varepsilon)} J_+ + \left(P_+(\pi) - \frac{\pi}{2(a-\varepsilon)} J_+ \right) e^{2(a-\varepsilon)(\theta-\pi)/\pi}. \tag{26}$$

Of course, when $\varepsilon = 0$, we recover the results obtained previously via the first profile. The value of J_- is as before (see eq. (21)), and

$$J_+ = \frac{2(a-\varepsilon)}{\pi} \frac{e^{-a}}{1 - e^{-(a-\varepsilon)} + (1-\varepsilon/a)} [P_+(0) - P_+(\pi)e^{\varepsilon}]. \tag{27}$$

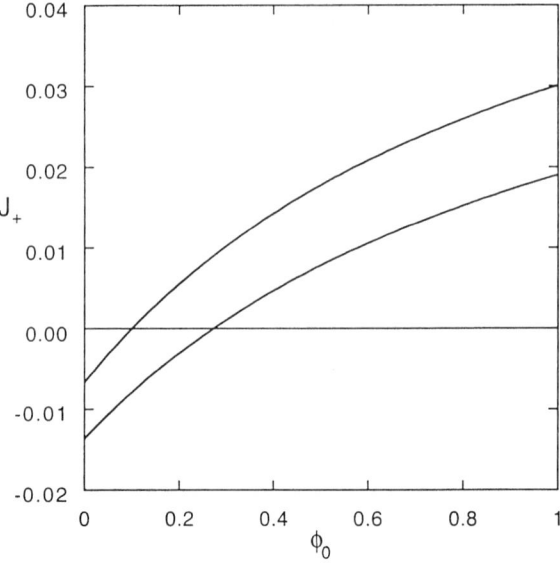

Figure 7.4. Flux as a function of external substrate concentration. The top curve is for the topmost profile of fig 7.2b; the bottom curve uses the bottom-most profile with $\varepsilon = 1$. Also, $\phi_\pi = 0.1$ in both cases. See text for details.

Full expressions for $P_-(0)$, $P_-(\pi)$, $P_+(\pi)$, and $P_+(0)$ are rather cumbersome and are not given here. Now, if the intrinsic binding and unbinding rates are not affected when $\varepsilon > 0$, so that the transporter interacts in the same way with its substrate environment, then it is easy to show that transport shuts off when $\phi_\pi e^\varepsilon = \phi_0$, or when $\phi_0 > \phi_\pi$, as expected. Figure 7.4 compares the flux as a function of ϕ_0 when $\varepsilon = 0$ (top curve) and when $\varepsilon = 1$ (bottom curve). Here, $\phi_\pi = 0.1$, $a = 2$, $\gamma_+^0/\gamma_-^0 = \phi_0$, and $\gamma_+^\pi/\gamma_0^\pi = \phi_\pi$ for both curves. When both transporters are carrying substrate from the exterior to the cell interior for a given ϕ_0, the $\varepsilon = 0$ one is faster; conversely, when both transport from the interior out, the $\varepsilon = 1$ protein has the larger (more negative) flux.

We complete this section with a word about the piecewise linear potentials. It is clear that their use allows for analytic solution to the differential equations. Now, as we approach a corner—at $\theta = \pi/2$ in a given potential, for example—from the right and from the left, we have ensured that the values of the flux calculated from either direction are the same. That is, we set $J^{(1)}(\pi/2) = J^{(2)}(\pi/2)$. The values of the resulting probability densities are also equal by construction at the junction of the two underlying potential legs. However, it is easy to see that the *derivatives* of the densities cannot match at the corner: $^{(1)}(\theta)/d\theta \neq dP^{(2)}(\theta)/d\theta$ at $\theta = \pi/2$. As seen from eq. (2), if the fluxes and densities are to be nonsingular at the junction of the legs, the derivatives of the densities must be singular, since the derivatives of the *potentials* do not match. Nevertheless, because the physically important quantities are fluxes, the piecewise potentials provide for a safe and useful approximation.

We turn now to the thermodynamics of transport.

3. Thermodynamics of Rectified Conformational Diffusion

It is natural to write down a (dimensionless) chemical potential—which contains the thermodynamics of the problem—for the protein when it has conformation θ:

$$\Omega_+^{(1)}(\theta) = U_+^{(i)}(\theta) + \ln P_+^{(i)}(\theta), \tag{28}$$

$$\Omega_-^{(i)}(\theta) = U_-^{(i)}(\theta) + \ln P_-^{(i)}(\theta). \tag{29}$$

For the protein–substrate complex, as described via eq. (28), the index i equals 1 or 2, referring to the first and second legs of the underlying potential of mean force. The free protein, with chemical potential given by eq. (29), will have i equal to 3 or 4. In essence, we have proceeded in the usual spirit of statistical mechanics: given a potential energy (here, a potential of mean force) we find a probability density which, in turn, yields a free energy. Now, the derivative of the chemical potential will determine, as we discussed in the introduction, the velocity of conformational changes when the protein has structure θ:

$$v_+^{(i)}(\theta) = -\frac{d\Omega_+^{(i)}(\theta)}{d\theta}, \tag{30}$$

$$v_-^{(i)}(\theta) = \frac{d\Omega_-^{(i)}(\theta)}{d\theta}, \tag{31}$$

for velocities in the unbound spaces. Indeed, eqs. (30) and (31) follow simply from the flux equations for conformational diffusion (see eq. (2)) via division by the probability density:

$$\frac{J_+(\theta)}{P_+(\theta)} = \frac{d[U_+(\theta) + \ln P_+(\theta)]}{d\theta}, \tag{32}$$

where we have omitted the superscript. There is, of course, a similar expression for the unbound protein. It is clear that $v \equiv J/P$. (While this work was in progress, we learned that Doi and Edwards (1986) develop in the same way a chemical potential for a Brownian particle.)

Note that the conformational fluxes are decoupled, and that this has been exploited in the above treatment: for $0 \leq \theta \leq \pi$, say, the protein is loaded, which leads to $v_+(\theta)$ through eq. (32) in this range of θ. Also, the velocity is defined at some conformation, but the experimentally observed transport rate is obtained by ensemble averaging over all conformations. That is,

$$\text{transport rate} \equiv \langle v(\theta) \rangle = \int v(\theta) P(\theta) d\theta = J \int d\theta,$$

since we work in steady state. The transport velocity is proportional to the flux, as was stated previously.

As discussed in the introduction, in so far as the Fokker–Planck equation can capture the elements of conformation diffusion, the development above should be independent of substrate concentration gradient. When a protein binds substrate and starts on its conformational path to transport its load, the protein–substrate complex is no longer affected by the presence of substrate, since it cannot bind another molecule. The biopolymer's fluctuations are caused by thermal noise and not the amount of substrate (unless the substrate gradient affects in some way the features of the noise). Similarly, the fluctuating free protein diffuses through the same mechanism and with the same dynamics independent of the gradient. However, the kinetics of substrate binding depend on concentration, but this changes the values of $P_-^{(i)}(\theta)$ and $P_+^{(i)}(\theta)$ at θ and not their forms or the form of the chemical potential.

Return again to the functioning protein where transport takes place from the cell's exterior to the interior. As already shown, the bound protein will diffuse unidirectionally from $\theta = 0$ to $\theta = \pi$; the free protein will travel from $\theta = \pi$ to $\theta = 2\pi (\equiv 0)$. From the *thermodynamic* view, the bound protein must flow downhill in chemical potential in the range $0 \leq \theta \leq \pi$—$d\Omega_+^{(i)}(\theta)/d\theta < 0$ here—while the unloaded protein is characterized by $d\Omega_-^{(i)}(\theta)/d\theta < 0$ for $\pi \leq \theta \leq 2\pi$. This is shown in fig. 7.5, where $\Omega(\theta)$ labels the chemical-potential legs in eqs. (28) and (29), and it is understood that, for the ranges $0 \leq \theta \leq \pi/2$, $\pi/2 \leq \theta \leq \pi$, $\pi \leq \theta \leq 3\pi/2$, and $3\pi/2 \leq \theta \leq 2\pi$, we plot, respectively, $\Omega_+^{(1)}(\theta)$, $\Omega_+^{(2)}(\theta)$, $\Omega_-^{(3)}(\theta)$, and $\Omega_-^{(4)}(\theta)$. We have used here the topmost potential $U_+(\theta)$ from fig. 7.2b with $a = 2$. The relevant equations to employ in conjunction with eqs. (28) and (29) are eqs. (16)–(25). The substrate concentration in the cell is fixed ($\phi_\pi = 0.1$) with $\phi_0 = 0.1, 0.2, 0.5,$ and 0.9. As the external concentration ϕ_0 increases, the slope of the chemical potential, both for the bound and free protein, becomes more negative corresponding to more rapid transport. Clearly, when $\phi_0 = \phi_\pi = 0.1$, the velocity $v(\theta)$ is zero. If $\phi_0 < \phi_\pi$, then $d\Omega_+/d\theta$ and $d\Omega_-/d\theta$ would be positive, indicative of net transport from the cell interior to the outside (this case is not shown).

What is the meaning of the "jump" at $\theta = \pi$ in going from the bound to the unbound state? We have used $\gamma_+^\pi/\gamma_-^\pi = \phi_\pi$ (and $\gamma_+^0/\gamma_-^0 = \phi_0$), and since $\phi_\pi < 1$, it is effectively easier for the protein to *unbind* at $\theta = \pi$. (Likewise it is easier for the protein to unbind at $\theta = 0$). Hence, since $P_+(\pi) = \gamma_+^\pi P_-(\pi)/\gamma_-^\pi$ from eq. (15), $P_+(\pi) < P_-(\pi)$ and, therefore, $\Omega_+^{(2)}(\pi) < \Omega_-^{(3)}(\pi)$. Now, the flux in one space—from which we identify the chemical potential as

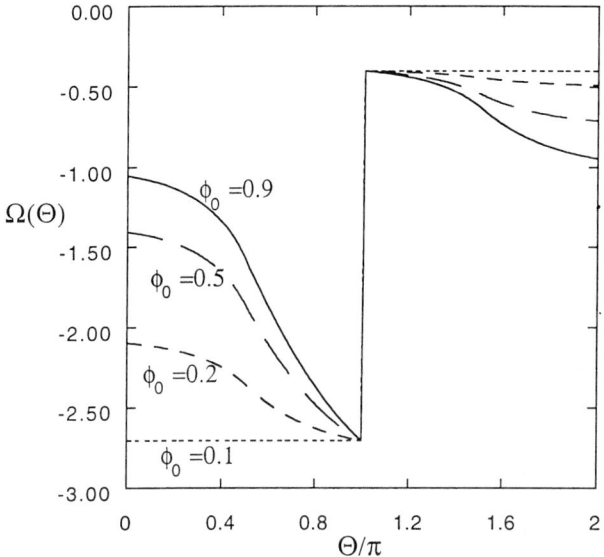

Figure 7.5. Chemical potential vs. conformational "position" of the transporter. For $0 \leq \theta \leq \pi$, the transporter is bound; for $\pi \leq \theta \leq 2\pi$, it is free. In each case shown, $\phi_\pi = 0.1$ with $\gamma_+^0/\gamma_-^0 = \phi_0$ and $\gamma_+^\pi/\gamma_-^\pi = \phi_\pi$, which is also true for the remaining figures.

in eq. (32)—is decoupled from that in the other, and hopping between spaces is not explicitly considered when we write down the flux. In other words, we have characterized the thermodynamic driving force for conformational change in one space, but we do not capture the thermodynamics of the transformation between spaces. The transporter is *not* going uphill at $\theta = \pi$: this "jump" is an "artifact" from the nature of our chemical potentials which describe the flow within one space only and not between spaces. $\Omega_+^{(2)}(\pi) < \Omega_-^{(3)}(\pi)$ simply indicates that there is higher probability for the transporter to be unbound at $\theta = \pi$.

It is easily seen that the chemical–potential legs in a given space, though constructed from potentials of mean force which join in a corner, have no singularity (and none in their derivatives) at the matching point. Figure 7.6 shows the effect of increasing barrier height a when $\phi_\pi = 0.1$ and $\phi_0 = 0.9$. We have used here again the top-most bound potential, and note that the velocity slows as the chemical potentials flatten with increasing barrier height. For a given large barrier height, the velocity of conformational change remains appreciable only at the top of the barrier, because it is here that the probability density is smallest, and $v \equiv J/P$ (with the flux constant). In effect, as a increases, we are uncovering the thermodynamics of a rate-theory formulation for the problem. Previously, it was stated that using the middle potential shown in fig. 7.2b (the one with the well at $\theta =$

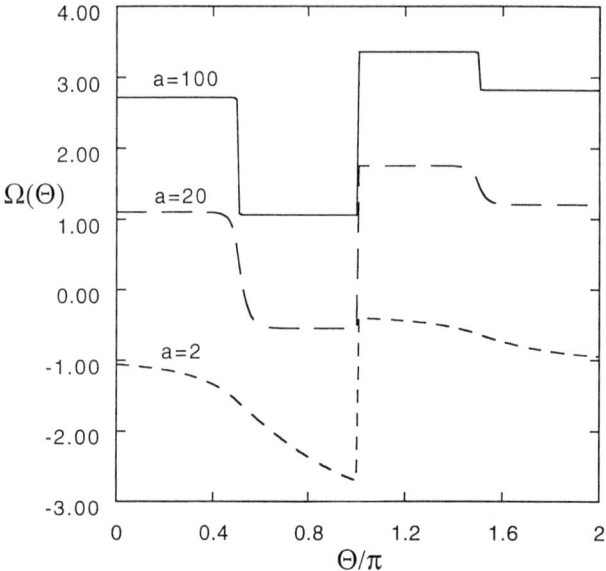

Figure 7.6. Effect of increasing barrier height a on chemical potential, with $\phi_\pi = 0.1$ and $\phi_0 = 0.9$ for each curve.

$\pi/2$) leads to the same flux as that obtained from the topmost profile. At the level of the chemical potential, however, which is a "local" quantity dependent on the particular conformation, we see differences between the results arising from the two profiles. Figure 7.7 depicts these differences when $\phi_\pi = 0.1$ and $\phi_0 = 0.9$ for both cases, with $a = 2$.

Though it is not necessary to show a figure, it is clear that using a bound potential with $U_+(0) < U_+(\pi)$ gives a downhill chemical potential path from $\theta = 0$ to $\theta = \pi$ when transport is from the otuside to the inside of the cell: though the *mechanism* dictates that the protein–substrate complex travels to higher potential of mean force, the *thermodynamic* flow is still downhill. Once again, this thermodynamic path arises from the coupling of the conformational dynamics to the substrate concentration gradient.

4. Discussion

Our treatment of protein dynamics relies upon the existence of a well-defined conformational pathway which is characterized abstractly through the coordinate θ. Unraveling the true nature of this pathway is a formidable task, the final result being an understanding of the atomic-level motions leading to protein function and the projection of this detailed information onto a single coordinate. As discussed in section 1, computer

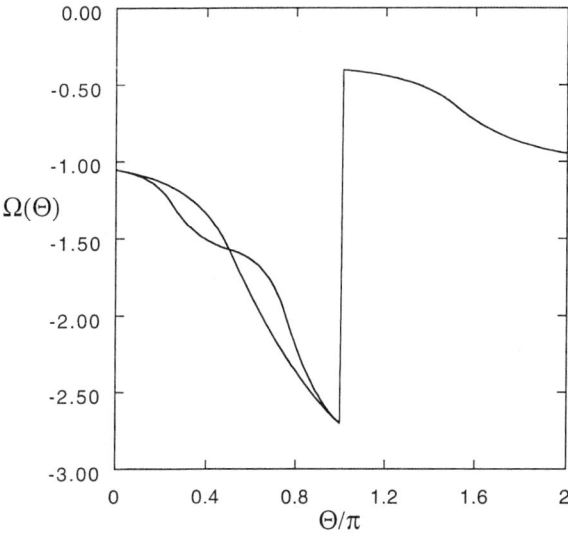

Figure 7.7. Comparison of chemical potentials arising from the topmost and middle potential profiles of Fig. 2b. Here $a = 2$, $\phi_\pi = 0.1$, and $\phi_0 = 0.9$. The curve with more structure corresponds to the middle potential.

cimulations shed light on fluctuations in the picosecond timescale. Now, such simulations often evolve from a consideration of the Langevin dynamics of each atom (Nadler et al., 1987). In contrast, our work proceeds from a Langevin (or the equivalent Fokker–Planck) equation in the single abstract conformational coordinate, but we are not the first to describe a complex system via a simplified "reaction" coordinate. For example, electronic relaxation of dye molecules in solution has been coupled to molecular conformations modelled by a single pathway (Bagchi et al., 1983), and electron-transfer reactions induced by (or coupled to) solvational and intramolecular vibrational motion has been described by a FPE in the solvent polarization coordinate (Sumi and Marcus, 1986; Nadler and Marcus, 1987). Zwanzig (1992) has discussed the kinetics of ligand binding to myoglobin in terms of a Langevin description for the radius of the fluctuation bottleneck leading to the binding site in the protein's interior, while de Gennes (1975) used a single coordinate to locate the boundary of the helix–coil transition in a polymer. Indeed, the intricate process of protein folding (Frauenfelder and Wolynes, 1994) is often caricatured as a protein molecule traveling on a free-energy landscape—described by one coordinate—and funneled to the global minimum. The work of Shaitan and Rubin (1982), in which they coupled electron transport with conformational mobility of a protein, is more in spirit with our studies (at least in mathematical structure): the fluctuating protein accepts an electron at some preferred conformation, diffuses along a new potential surface, and donates the electron from a preferred coordinate on this new surface. Nonetheless, the direction for flow of conformational probability density cannot be reversed in their picture. In a numerical study, but beginning again from a Fokker–Planck formulation with parabolic potentials, Cartling (1985) has investigated the transient evolution of the probability density for the coupled protein conformation and electron transport problem. In another work, using two parabolic potentials to describe enzymatic states separated by a reversible chemical reaction (like substrate binding), Kurzynski (1994) has mapped the FPEs to an equivalent Schrödinger picture. A single FPE that describes only the loss of conformational probability density from a potential surface has been used to examine the binding kinetics of ligand to heme proteins (Agmon and Hopfield, 1983). More recently, a quantum-mechanical treatment of the binding problem has been provided by Schwartz (1991).

We stress again that our work addresses the entire cycle of protein activity, with substrate binding and unbinding to special protein conformations. When the protein is loaded with substrate, the rectified fluctuations of the protein–substrate complex serve to transport the load across the membrane—this bound fluctuation leg defines the function of the transporter. The subsequent directional diffusion of the unloaded protein "restores" the biopolymer to the form which restarts the cycle. The scheme we have adopted allows for connection with rate theories of enzymatic activity, since these implicitly assume as the active species a particular conformation of the enzyme. Furthermore, our statistical-mechanical avenue leads in a natural way to the thermodynamics—via a force-flux equation for conformational change—of a class of Brownian molecular motors, in this case, the membrane transport proteins. It is intriguing to consider such an analysis for the Brownian machines discussed in the introduction.

Our "microscopic" approach—microscopic is in quotes because the potential surfaces on which the protein diffuses are not at present available from any fundamental calculations—opens the way for investigation of non-delta-function binding of the substrate. That is, what can be said about the dynamics of transport when the substrate interacts with a spread of closely related protein conformations so that the bound and unbound fluctuation spaces are not decoupled?

Finally, can we understand how ATP interacts with a transporter so that the substrate is carried up the concentration gradient? We would expect ATP binding and hydrolysis to affect the potential profiles or the substrate binding constants to the protein, and such considerations can be incorporated in our analysis.

Acknowledgments We have enjoyed and have learned from discussions with Frank Novak, Martin Bier, George Oster, and Marcelo

Magnasco. CKB is grateful to David Oxtoby for helpful input and financial support during the course of this work. RDA is supported through NIH Grant No. R01ES06010.

References

Agmon, N., and Hopfield, J. J. (1983) *J. Chem. Phys.* **78**, 6947; **79**, 2042.
Astumian, R. D., (1997) *Science* **276**, 917.
Bagchi, B., Fleming, G. R., and Oxtoby, D. W. (1983) *J. Chem. Phys.* **78**, 7375.
Bagdassarian, C. K., and Oxtoby, D. W. (1994) *J. Chem. Phys.* **100**, 2139.
Careri, G. (1974) In: *Quantum Statistical Mechanics in the Natural Sciences*, S. Mintz and S. Widmayer (eds.). Coral Gables: Plenum.
Cartling, B. (1985) *J. Chem. Phys.* **83**, 5231.
Darnell, J., Lodish, H., and Baltimore, D. (1990), *Molecular Cell Biology*, New York: Scientific American Books.
DeGennes, P. G. (1975) *J. Stat. Phys.* **12**, 463.
De Groot, S. R., and Mazur, P. (1984) *Non-Equilibrium Thermodynamics*. New York: Dover.
Doi, M., and Edwards, S. F. (1986) *The Theory of Polymer Dynamics*. Oxford: Oxford University Press.
Frauenfelder, H., and Wolynes, P. G. (1994) *Physics Today* **47**, 58.
Hanggi, P., and Bartussek, R. (1996) In: *Nonlinear Physics of Complex Systems – Current Status and Future Trends*, J. Parisi, S. C. Muller, and W. Zimmerman (eds.). Berlin: Springer.
Jackson, M. B. (1993) *J. Chem. Phys.* **99**, 7253.
Julicher, F., Ajdari, A., and Prost, J. (1997) *Rev. Mod. Phys.* **69**, 1269.
Karplus, M., and McCammon, J. A. (1986) *Sci. Am.* **254**, 42.
Kurzynski, M. (1994) *J. Chem. Phys.* **101**, 255.
McCammon, J. A., and Harvey, S. C. (1987) *Dynamics of Proteins and Nucleic Acids*. Cambridge: Cambridge University Press.
Nadler, W., Brünger, A. T., Schulten, K., and Karplus, M. (1987), *Proc. Natl. Acad. Sci. USA* **84**, 7933.
Nadler, W., and Marcus, R. A. (1987) *J. Chem. Phys.* **86**, 3906.
Schwartz, S. D. (1991) *Chem. Phys. Lett.* **185**, 16.
Serra, R., Andretta, M., Compiani, M., and Zanarini, G. (1986) *Introduction to the Physics of Complex Systems*. Oxford: Pergamon.
Shaitan, K. V., and Rubin, A. V. (1982) *Mol. Biol. (USSR)* **16**, 1004.
Sumi, H., and Marcus, R. A. (1986) *J. Chem. Phys.* **84**, 4894.
Svoboda, K., Schmidt, C. F., Schnapp, S. J., and Block, S. M. (1993) *Nature* **365**, 721.
Zwanzig, R. (1992) *J. Chem. Phys.* **97**, 3587.
Zwanzig, R., Szabo, A., and Bagchi, B. (1992) *Proc. Natl. Acad. Sci. USA* **89**, 20.

Index

accessible surface area, 21, 22
affinity
   estimation by simulation, 90–93
   free-energy perturbation, 90–93
Ala scanning mutagenesis, 50
AMBER, 99–100, 105
AMSOL, 104–105
analytical continuum electrostatics (ACE), 17, 23
association, 126–127
atomic solvation parameters, 20

binding energy, 124
Brownian dynamics, 126
Brownian ratchet, 205, 224
   and molecular motors, 205
   and translational motion, 205

cavities, 38
charge distribution, 114–115
   explicit, 114
   mobile, 115
chemical modification, 181–188, 193–198
chemical potential, 204, 219–222
CoMFA, 93, 102
comparative sequence analysis, 181, 191–193
conformational fluctuation, 206–211
cooperativity, 49, 59
   in substrate recognition, 67

cytochrome p450, 90, 92

dielectric constant, 114
   solvent, 115
   protein, 115
   saturation, 115
dielectric pressure, 118
double-mutant cycles, 56
dynamic-programming algorithm, 183–187, 189–190

electron transfer, 123
enthalpy, 26, 117, 124–126
   of unfolding, 26
entropy, 88–99, 117, 124–126
   configurational, 12
   cratic, 88
   estimation by heuristics, 96–99
   estimation by simulation, 90–93
   internal, 96–97
enzyme specificity, 59
epitope, 50

fibrinogen, 52
Field, Maxwell, 113
Fokker–Planck equation, 211–218
force-flux equation, 204, 219–221
free energy, 204, 219–222
free energy decomposition, 33–40

generalized Born model, 23

INDEX

genetic algorithm, 190
Gibbs free-energy, 180
GOLPE, 105–106

Hamiltonian, 9
HASL, 93
heat capacity, 31–33, 124–124
  of unfolding, 31–33
heuristic approaches, 93–103
HIV protease, 91–94, 102, 104, 106–107
HOMO, 104
hydrogen bond, 36, 39, 105–106
  contribution to protein stability, 36, 39

LUDI, 94–95, 99

Kirkwood superposition approximation, 41

MacroModel, 99–100, 104–105
Mix&match, 192, 200–202

nearest-neighbor, 181–182
neural networks, 101–102
  theoretical considerations, 87–90
  VALIDATE predictions, 101–103
nucleic acids, 135–180
  enthalpy of melting, 140–142
  ionic strength dependence of melting, 164
  loops and bulges, 165–168
  melting of double helices, 143, 169
  melting temperature, 136
  mismatches, 165
  nearest-neighbor parameters, 161–163
  nuclear magnetic resonance, 151, 152
  partition functions, 147–151
  secondary structure definition, 157
  secondary structure prediction, 170–172
  temperature jump, 144, 145
  two-state equilibria, 138

partial least squares of latent variables (PLS), 96, 101–102, 105–106
partition coefficient (logP), 98–99, 104
phospholipase, 96
pK_a, 120–122
Poisson–Boltzmann, 22, 29, 113–126
  model, 113
  equation, 116
  solutions to, 119
  use with brownian dynamics, 126
polarizability, 114–118
  electronic, 114, 118
  dipolar, 118
potential, 115–117
  average, 115
  coulombic, 116
  screening, 116
  reaction, 117
proton release, 122

QM/MM, 16
QSAR 3D, 93–94

reaction coordinate, 204, 208
redox midpoint potential, 120, 123
R2 retrotransposon, 190–197
RNA secondary structure, 177–202
  accuracy of prediction, 185
  chemical probes, 187–188
  constraining prediction, 187–188, 196–198
  definition of, 178–180
  enzymatic probes, 187
  equilibrium, 180
  formation, 180
  partition function, 189

sequence alignment, 177, 191–193
serine proteases, 59
solvation, 17–25, 101–104, 116–119
  empirical models, 20
  energy, 119
  Gaussian exclusion model, 25
  GB/SA model, 101, 104

INDEX   229

potential, 116
statistical thermodynamics of, 17
surface estimates, 100–101
SONNIC, 102

T4 lysozyme, 90
target-primed reverse
    transcription, 191
thermodynamic cycle, 27, 36

thermodynamics, 5–7
    irreversible, 5
    second law of, 3, 4
    vs kinetics, 7
thrombin, 52, 60
time scales, 206–207

VALIDATE, 97–103
VALIDATE II, 103–107